Neues verkehrswissenschaftliches Journal

Ausgabe 22

Hybrid Model for Proactive Dispatching of Railway Operation under the Consideration of Random Disturbances in Dynamic Circumstances

Von der Fakultät Bau- und Umweltingenieurwissenschaften der Universität Stuttgart
zur Erlangung der Würde einer Doktor-Ingenieurin (Dr.-Ing.)
genehmigte Abhandlung

Vorgelegt von

Weiting Zhao

aus Shaanxi, VR China

Hauptberichter:	Prof. Dr.-Ing. Ullrich Martin
Mitberichter:	Prof. Dr.-Ing. Rainer König

Tag der mündlichen Prüfung: 21.09.2017

Institut für Eisenbahn- und Verkehrswesen der Universität Stuttgart

2017

Titelbild: Weiting Zhao

Herstellung und Verlag: Books on Demand GmbH, Norderstedt

Printed in Germany

ISBN 978-3-7460-3501-7

Preface

Railway operations are regularly subject to stochastic influences that cannot always be compensated for by the recovery times and buffer times contained in the timetable, and can thus result in passenger delays. For this reason, railway dispatching is of particular importance because it helps to ensure on-time railway operations.

The most significant result of this dissertation is a realistic dispatching algorithm that can be used in both real-life applications and railway simulations. The algorithm is able to generate dispatching solutions under consideration of stochastic disturbances with consideration of different risk levels of the individual elements of train paths or infrastructure elements; the risk levels are quantified using the negative influences caused by the operational disturbances. The knowledge gained in this dissertation on the modeling and methodical development of a dispatching algorithm is a significant scientific contribution to the efficient management of railway operations in practice and in railway operation simulations directly relevant to practical implementation.

Stuttgart, September 2017

Ullrich Martin

Acknowledgement

This dissertation summarizes my scientific research at Institute of Railway and Transportation Engineering in the University of Stuttgart.

First of all, I would like to express my deep gratitude to my supervisor Prof. Dr-Ing. Ullrich Martin for accepting me and giving me the opportunity to do the research of my interested field. His sagacious guidance and remarkable suggestions during the four years made me understand my research field more intensively. Without constant discussion with him and technical support, advice from him during the entire working period of my dissertation, I can hardly complete the dissertation. His visionary thoughts, instructive notes, and constructive hints are critical to enhance my knowledge and experience. Special thanks to my second supervisor Prof. Dr.-Ing. Rainer König for examining my dissertation.

I am also very grateful to my colleague Dr.-Ing. Yong Cui. He inspired me a lot during the process of dynamic dispatching algorithm developments. The challenging and fruitful discussions frequently with him made me precisely understand the whole theory system of railway operation and simulation. I am also indebted to all of the colleagues in the institute. Many thanks to my colleagues Dr.-Ing. Zifu Chu, Dr.-Ing. Xiaojun Li and Dr.-Ing. Jiajian Liang. They offered me valuable suggestions related to my dissertation topic and help modifying the manuscript of the project proposal. Mr. Syed Murtaza Hasan provides strong support for the English polishing of this dissertation. Dipl.-Vw.tech. Carlo von Molo helped me a lot with the template, posters and all the other things regarding to organization as well as format styles. Dipl.-Inf. Stefan Schmidhäuser provided me all the accessibility to the necessary software as well as the usable account for all the meaningful resources. I am also deeply thankful to my parent for encouraging and supporting my interests and respecting my determination. And special thanks to all my friends who accompany me every wonderful day.

Thanks to all the people who positively influence the accomplishment of my dissertation.

Content

List of Figures

List of Tables

Abstract

With the increasing traffic demand and limited infrastructure expansion, railway networks are often operated close to the full capacity, especially in heavily used areas. As a result, the basic timetable is quite susceptible to the operational disturbances, and thereby the propagation and accumulation of delays significantly degrade the service level for customers. To solve this problem, extensive researches have been conducted by focusing on the predefined robust timetables and the real time dispatching algorithm development. However, it has been widely recognized that excessive robust timetables may deteriorate the operating capacity of the railway network and the addition of recovery time and buffer time can be hardly implemented in the congested area. Moreover, most of the conventional dispatching algorithms ignore the further potential random disturbances during the dispatching process, which yield non-implementable dispatching solutions and, as consequences, inferior punctuality and repetitive dispatching actions. To this end, this project aims to develop a new algorithm for real-world dispatching process with the consideration of risk-oriented random disturbances in dynamic circumstances. In the procedure of this project, an operational risk map will be firstly produced: by simulating considerable amount of disturbed timetables with random disturbances generated in a Monte-Carlo scheme and calculating the corresponding expected negative impacts (average total weighted waiting time among all the disturbances scenarios), different levels of operational risk will be assigned to each block section in the studied railway network. Within a rolling time horizon framework, conflicts are detected with the inclusion of risk-oriented random disturbances in each block section, and the near-optimal dispatching solutions are calculated by using Tabu search algorithm. Finally, three indicators including total weighted waiting time, the number of relative reordering and average absolute retiming, are chosen to testify the effectiveness and advantages of the proposed dispatching algorithm. The sensitivity analysis of dispatching-related parameters is conducted afterwards to investigate the most appropriate relevant parameters for the specific studied area. The proposed algorithms are expected to be capable of automatically producing near-optimal and robust dispatching solutions with sufficient punctuality achieved.

Zusammenfassung

Aufgrund des steigenden Verkehrsaufkommens und der begrenzten Erweiterbarkeit der Infrastruktur werden Eisenbahnnetze, insbesondere in hoch belasteten Bereichen, häufig an ihrer Belastungsgrenze betrieben. Hieraus resultiert, dass Basisfahrpläne anfällig gegenüber Betriebsstörungen sind, und die daraus entstehenden Folgeverspätungen die Betriebsqualität verschlechtern. Zur Lösung dieses Problems wurden bereits umfangreiche Forschungsarbeiten zur Weiterentwicklung der festgelegten stabilen Fahrpläne und des Echtzeit-Dispositionsalgorithmus geleistet. Jedoch wurde erkannt, dass zu stabile Fahrpläne die Betriebskapazität des Eisenbahnnetzes inakzeptabel absenken können und Fahrzeitzuschläge sowie Pufferzeiten im überfüllten Netzabschnitt systematisch nur schwer implementierbar sind. Überdies beziehen konventionelle Dispositionsalgorithmen zukünftige zufallsbedingte Unsicherheiten nicht in den Dispositionsprozess ein. Hierdurch werden praxisgerechte Dispositionslösungen ausgeschlossen, woraus wiederum eine schlechtere Pünktlichkeit und iterative Dispositionshandlungen resultieren. Um dem entgegen zu treten, soll in diesem Projekt ein neuer Algorithmus für Echtzeit-Dispositionsprozesse entwickelt werden, der risikoorientiert zufallsbedingte Unsicherheiten im künftigen Betriebsablauf berücksichtigt. Im Verlauf dieses Projekts wird zunächst eine betriebliche Risikokarte erstellt: Durch die Simulation von gestörten Fahrplänen in großer Anzahl mit zufallsbedingten Unsicherheiten nach dem Monte-Carlo-Verfahren und der Berechnung der zugehörigen erwarteten negativen Auswirkungen (durchschnittliche gesamten gewichteten außerplanmäßigen Wartezeit unter allen Störungsszenarien) können jedem Blockabschnitt des betrachteten Eisenbahnnetzes verschiedene Risikostufen zugewiesen werden. Innerhalb eines rollierenden Zeitrahmens können Konflikte unter Einbeziehung von zufallsbedingten Unsicherheiten für jeden Blockabschnitt risikoorientiert entdeckt und die nahezu optimalen dispositiven Lösungen mithilfe des Tabu-Suchalgorithmus generiert werden. Schließlich werden drei Indikatoren, einschließlich der gesamten gewichteten außerplanmäßigen Wartezeit, der Anzahl der relativen Veränderung der Zugreihenfolge und des durchschnittlichen absoluten Abweichung zwischen dem Basisfahrplan und dem Dispositionsentscheidung ausgewählt, um die Wirksamkeit und die Vorteile des vorgeschlagenen Dispositionsalgorithmus nachzuweisen. Die Sensitivitätsanalyse der Dispositionsparameter wird anschließend durchgeführt, um die geeignetsten relevanten Parameter für den gewählten Untersuchungsraum festzulegen. Die vorgeschlagenen Algorithmen sollen in der Lage sein, automatisch die nahezu optimale Dispositionslösungen zu erzeugen, die auf stabilen Dispositionsfahrplänen mit hinreichender Pünktlichkeit beruhen, indem bereits bei der Generierung dieser Fahrpläne die gegenseitigen Wirkungen der Dispositionsmaßnahmen berücksichtigt werden.

1   Introduction

1.1   Background and motivation

Railway systems contain various endogenous and exogenous disturbances that may lead the train operation to deviate from the basic schedule. Due to the increasing traffic demand and limited infrastructure expansion, railway networks are often operated close to their full capacity, especially in heavily used areas or bottleneck sections. Thus, the schedules are more susceptible to even small operational disturbances[1] caused by technical devices failure, human behavior and external extreme environment. Propagation and accumulation of delays are triggered afterwards, leading to degradation of service level for customers. According to the statistical data from *Süddeutsche Zeitung* (SZ) train monitor, over 20% of train runs were delayed more than 15 minutes from April 2012 until March 2013 in Germany (Plöchinger, Jaschensky 2013).

In normal situation, small disturbances could be neutralized by the buffer time[2] and recovery time[3] set in the basic timetable[4]. In existing researches, stochastic approximate relationships between scheduled headways and consecutive delays are studied so as to adjust train timetables according to the expected consecutive delay effects in (Carey, Kwieciński 1994). A stochastic optimization model is developed in (Kroon et al. 2008) to allocate the recovery time and buffer time in a given timetable in order to improve the robustness against stochastic disturbances as much as possible and to minimize the need of modifying the given timetable. An algorithm to compute the propagation of initial delays over a periodic railway timetable is developed using max-plus algebra in (Goverde 2010). The stability of timetables and the influence of traffic volume, traffic heterogeneity, primary delays on used recovery time is investigated in (Lindfeldt 2015).

---

[1] Operational disturbances: collective name includes entry delays, dwell time extensions, running time extensions and departure time extensions.

[2] Buffer time: an extra time which is added to the minimum line headway to avoid the transmission of small delays (Pachl 2002).

[3] Recovery time: a time supplement that is added to the pure running time to enable a train to make up small delays (Pachl 2002).

[4] Basic timetable: a basic timetable includes a set of information with detailed train paths, defining several months in advance the train order and timing at crossings, junctions and platforms.

However, two problems have been widely recognized. On one hand, a robust timetable with redundant buffer time and recovery time may deteriorate the operating capacity of the railway network and it plays almost no role in the heavily congested area. On the other hand, significant disturbances occurred during the railway operation can't be absorbed by the basic timetable itself and generate inevitably delays. In this situation, dispatching is necessary to improve the punctuality of railway operation.

At present, dispatching are often conducted manually in practice, and the quality of dispatching solutions highly depends on the experience of dispatchers. Dispatchers take dispatching actions based on certain predefined principles and objectives. For example, express passenger trains have higher priority than other trains, faster operated passenger trains have higher priority than slower operated trains in the case of the same train level (DB NETZ AG 2010), the train operated on time has higher priority than a delayed train, etc. In most instances, however, manual dispatching only deals with local conflicts without considering the dispatching effect on the global railway network sufficiently.

In this context, various optimization algorithms with global objective functions have been adopted inside the framework of computer-based dispatching models. And yet, very limited attention has been paid on further potential random disturbances during the dispatching process. Random disturbances occur frequently and unpredictably in real railway networks, forming a dynamic environment with high uncertainty. Accordingly, three shortages of the existing dispatching approaches are identified in the following aspects:

1) <u>Conflicts detection</u>: By detecting conflicted track sections only based on current railway states in deterministic process without consideration of potential random disturbances, potential conflicts are likely to be missed during calculating the dispatching solutions. Thus, the complemental dispatching actions may not be implemented because the dispatching points have already passed.

2) <u>Robustness of dispatching solutions</u>: During searching for optimal or near-optimal dispatching solutions, ignoring random disturbances that could possibly occur after the dispatching is done will fail to achieve robust dispatching solutions, and incur in repetitive dispatching actions. Frequent retiming and reordering will not only increase the burden of dispatchers and train drivers, but also increase the possibility of further conflicts and train delays.

3) <u>Punctuality</u>: Due to the lack of prospective disturbances consideration during the process of conflicts detection, some dispatching solutions cannot be conducted in

time, e.g., trains can only overtake the front one at the next dispatching point. The punctuality of the whole railway network can be influenced.

Accordingly, this project aims to propose a new dispatching algorithm under the consideration of further potential random disturbances. It is a hybrid dispatching model which combines both heuristic and simulation approaches. The produced dispatching solutions are expected to be more stable and more robust than that derived from current dispatching methods. Meanwhile, it is able to achieve a balance between punctuality and capacity of the whole railway network.

1.2   Research Objectives

The objective of this research is to generate robust and resilient dispatching solutions which are able to handle forthcoming disturbances with an extraordinary system performance in dynamic circumstances. Further potential random disturbances are worth being taken into consideration during the dispatching process in the heavily used areas. However, too much consideration of future random disturbances would lead to falloff in capacity of the whole railway network. To this end, it requires a continuous balance between maintaining a high utilization rate and sufficient robustness to minimize the sensitivity to disturbances.

Within a rolling time horizon (Peeta, Mahmassani 1995) frame, the most appropriate dispatching solutions are designed for each sub time period and then regularly updated by refreshing the state of railway operation from the field. The proposed dispatching algorithm and the generated dispatching solutions will be evaluated through simulation of the railway operation using the available software RailSys® (RMCON Rail Management Consultants 2010).

By implementing the proposed dispatching algorithm, several advantages are assumed to be achieved:

1) <u>Prospective detection of conflicts</u>. Prospective detection of conflicts can be achieved by considering potential random disturbances. Based on that, the generated dispatching solutions are capable of avoiding severe delay propagation and accumulation. In addition, complemental dispatching actions can be made before it is too late.

2) <u>Robust dispatching solution</u>. With the inclusion of potential random disturbances into the optimization model as one of the constraints, the obtained rescheduled timetable has stronger resistance to the unexpected disturbances occurred in real-world railway operations. With that, frequently reordering and retiming are significantly reduced.

3) <u>Punctuality</u>. Iterative dispatching in a dynamic time scheme and the allowance of imposing flexible disturbances depend on train types, block sections and railway networks are able to achieve a high degree of punctuality.

1.3 Contributions

As a result, the main outputs of this dissertation are:

1) <u>Module of operational risk analysis,</u> including the automatic generation of disturbed timetable within a certain target block section, calculation of operational risk indicator (see Section 3.2.3) of entire investigation network from simulation tools and production of the infrastructure operational risk map.

2) <u>Module of dynamic dispatching algorithm,</u> including automatic selection of near-optimal train sequence based on Tabu search, generation of rescheduled timetables with constraints and update of the actual state of railway operation until the start of the next stage.

3) <u>Module of inter-comparison with other dispatching algorithms,</u> including evaluation of the system with three metrics between *First-Come First-Serve (FCFS)* principle and the proposed dynamic dispatching system.

4) <u>Module of sensitivity analysis of dispatching-related parameters,</u> including automatic conduction of sensitivity analysis and selection of the appropriate dispatching-related parameters for a specific investigation of railway networks.

The operational risk analysis is the preliminary study of developing the dynamic dispatching algorithm. It provides a theoretical basis for recognizing the vulnerable block sections, on which the occurred disturbances can hardly be neutralized by the scheduled timetable reserves. It is meaningful for timetabling (e.g. allocate the recovery time and buffer time regarding to different block sections), dynamic dispatching (e.g. consider potential forthcoming disturbances in the process of conflict detection), and infrastructure maintenance (e.g. determine the priority for the network segments construction or maintenance).

Meanwhile, the main output of this dissertation, the hybrid dynamic dispatching system, offers a scientific way for the operators to select the sufficient optimal dispatching solutions automatically based on the combination of simulation method and metaheuristic method. This model is also expected to shed light on the importance and complexity of taking future potential stochastic disturbances into account in the procedure of conflict detection phase.

The inter-comparison module testifies the applicability and the sensitivity analysis module helps configuration with implementation of the system.

In such a manner, the results are not only significant for the academic innovation but also practical for the railway service suppliers to ensure a robust and good performance for both passenger and freight transport. As such, this dissertation can be of interest to a broad readership including those interested in railway planning and optimization, for instance, optimization of capacity, design of timetables as well as dynamic dispatching with operational risk taken into consideration.

1.4 Methodology and outline of the thesis

An integrated knowledge in the field of train dispatching, operations control, simulation of railway and computer programming has been applied in this dissertation. A hybrid model which combines the simulation method and metaheuristic method is developed. It makes the model share the advantages from both of the two methods.

In fact, as the foundation of the methodology, simulation models play a significant role of studying the relationship between the infrastructure, operation, and rolling stock, so as to precisely depicting the railway operation process. Compared to the physical model (laboratory experiments of a specific part of railway network) or the analytical model (mathematical model based on queue theory), a simulation model provides an efficient medium for the investigation of large complex railway networks in an inexpensive way. The physical model is not only money-consuming but also fails to represent the railway operation with abundant flexibility in different scales of railway network. As for the analytical model, the investigation area, which can only be a simple small railway network or line segments (otherwise no solutions will be given due to the high computational complexity), immensely limits the widely application of this approach. For instance, there is no existing analytical model to accurately calculate the unscheduled waiting time due to the deceleration or acceleration process of a hindered train waiting for the clearance of the next block section and the consecutive delays which may cause for other trains.

Therefore, in order to quantitatively assess the disturbances-caused negative impacts and the propagation of those negative impacts (see Chapter 3), it is indispensable to apply simulation models to provide a proxy field of the railway operation. In this dissertation, the simulation model is not only applied for the offline evaluation (operational risk analysis module) but also used for the online optimization (dynamic dispatching module). During the online optimization process, the metaheuristic method (Tabu Search) is employed simultaneously. Even though the globally optimal solution is not guaranteed through a metaheuristic method, the computation efforts for finding the near-optimal solutions (efficiently explore the solution domain) is dramatically reduced. Moreover, compared to the metaheuristic such as Greedy algorithm, Tabu Search provides a way to avoid getting stuck in the local optimal solution by introducing a Tabu list, which to some extent ensures the quality of the dispatching solutions.

Last but not least, a rolling time horizon approach is adopted for the dispatching algorithm development in the proposed study. It was firstly proposed by (Peeta, Mahmassani 1995) and have been widely used in related research such as signal control, disruption management, timetable scheduling, aircraft landing problem, etc. (Aboudolas et al. 2010; Meng, Zhou 2011; Nielsen et al. 2012; Caimi et al. 2012; Girish 2016). The term "rolling" implies "overlap" between two consecutive stages systematically. Since a potential conflict is detected according to the estimated blocking time stairway of each train run and the random disturbances may occur at any time, it is necessary to check the state of railway operation regularly and adjust the predictive disturbances based on operational risk level. The quality of dispatching solutions generated for each stage then can be guaranteed as time goes by.

The outline of this dissertation is shown in Figure 1-1. Chapter 2 presents the overview of the state of the art in the fields of railway dispatching from the perspectives of disturbance distributions, robustness of railway operation, objectives of dispatching algorithms, dispatching models and the corresponding assistance software.

In the following Chapters 3-6, the integrated dispatching algorithm is explained in detail. The operational risk analysis module is explicitly illuminated in Chapter 3, including the categories of disturbances considered, the probability distributions applied as well as the selected indicators of operational risk. Afterwards, the workflow of the dynamic dispatching algorithm is demonstrated in Chapter 4. Four tasks are specified orderly within the framework of rolling time horizon. The way of taking operational risk into account, the considered dispatching strategies, the iterative process of updating the railway operation are stressed generally. In order to testify the proposed dispatching algorithm, three indicators are chosen for the evaluation in Chapter 5. Sensitivity analysis regarding to the dispatching-related parameters is conducted Chapter 6.

After a theoretical description of the developed algorithm, a case study on a reference example of the railway network is presented, so as to provide the audience an intuitive understanding. The corresponding results of Chapters 3-6 are analyzed in Chapter 7. Meanwhile, the specification about application of the proposed algorithm automatically is given in Chapter 8. Finally, the major conclusions and prospective of further research are summarized in Chapter 9.

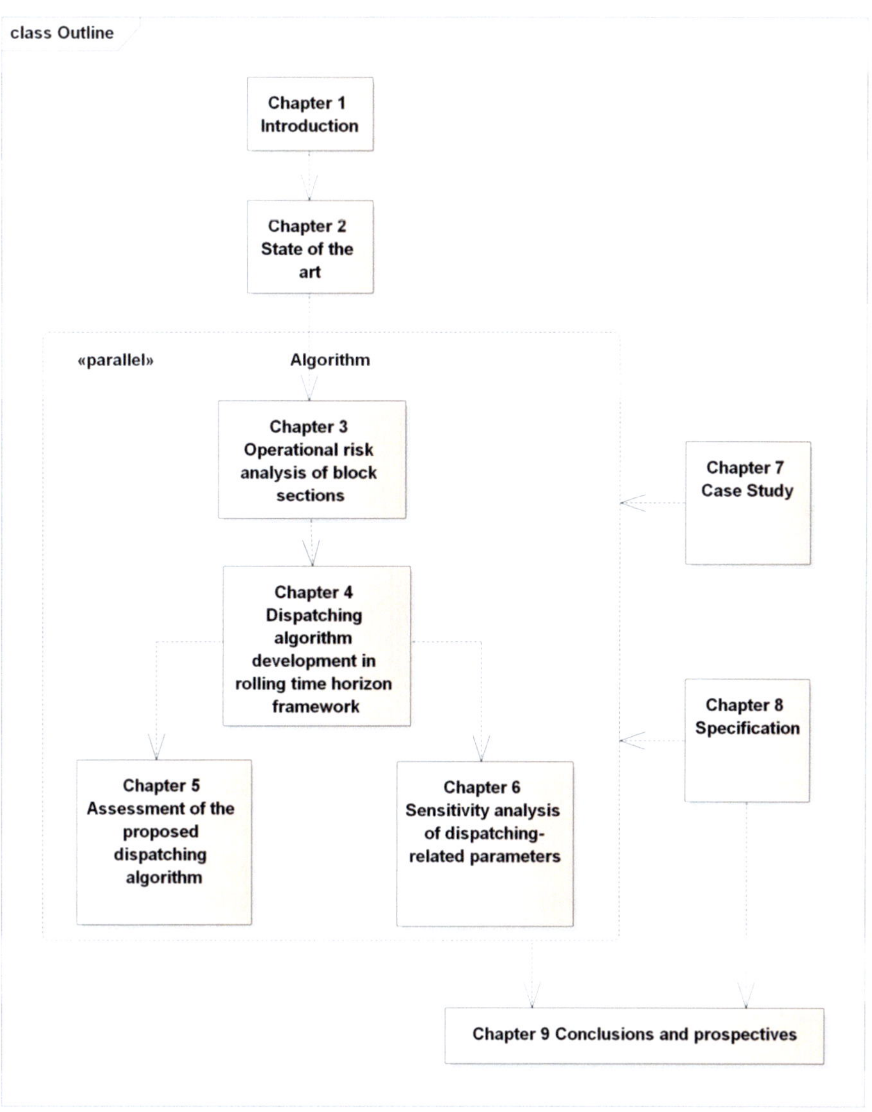

Figure 1-1   Outline of this dissertation

2 State of the art

2.1 Distributions of disturbances and delays

Served as a key indicator for evaluating the railway service quality, the delay of trains (measured by the unit of time in the manner of time extension) has drawn much attention of train service providers as well as academic researchers over decades. It is determined by the deviations between the actual and the scheduled arrival/departure/passing time at certain measurement points (Cui et al. 2016). The measured delays are composed of the primary delay (including the initial delay and the original delay) and the consecutive delay (Hansen, Pachl 2014).

However, it is labor-intensive to directly obtain delays at all measurement points from track occupation and release records. Therefore, large amount of models have been built to estimate the distribution of consecutive delays, arrival delays and departure delays based on the input distribution of the primary delays (often assumed based on limited empirical data). Schwanhäußer (1974) initially claimed that the pattern of arrival delays of passenger trains at a given station followed a negative exponential distribution, a statement which was later corroborated by (Goverde et al. 2001; Yuan et al. 2002; Goverde 2005). In further research, models such as the Weibull distribution, gamma distribution and lognormal distribution provided improved estimations of both arrival as well as departure delays (Bruinsma et al. 1999; Yuan 2006, 2006). In a review research, various distribution models were compared by Yuan (2006) in order to investigate how appropriately they conformed to train event times and processing times. This comparison was executed by utilizing the data collected at the Hague railway station. This comparison was executed by utilizing the data collected at the Hague railway station. Furthermore, Higgins and Kozan (1998) fit, in their research, a normal-lognormal mixed distribution to estimate the running times of trains between two stations using the data obtained for the German railway. Briggs and Beck (2007) found that the distribution of train delays can be described by so-called q-exponential functions for British railway network. In spite of mass studies on distributions of delays and delays propagation, it is still unrealistic to apply a standard distribution type which can be fit for all the railway networks. Because the distribution types of delays as well as its corresponding parameters vary from different track layouts, signaling systems, timetable design as well as train types.

During the railway operation process, there are two different kinds of perturbations which may cause delays and are more or less inevitable. One is the major perturbations, often known as disruptions, which are caused by broken down rolling stock, faulted overhead wires, etc., can be only handling with special disruption management strategies (e.g., rerouting, cancellation of trains or providing alternative connections with other transportation modes) (Nielsen et al. 2012). Under the circumstance of disruptions, severe negative impacts such as collapse of the entire railway network can be noticed. Maintaining the train service and keeping the connections between origins and destinations is the main concern for service providers. The other is the minor perturbations, often called as disturbances, whose distribution is often assumed based on experiences or calibration results with limited empirical data. It can be caused by bad weather conditions, stochastic driving patterns and prolonged boarding or alighting time of passengers. Disturbances can be either absorbed by the reserves (recovery time and buffer time) which previously set in the timetable or compensated by small dispatching actions (e.g., retiming or reordering) in real operation. In order to improve the railway operation stability and maintain the railway operation robustness, studies of disturbances distributions are necessary. Up to now, limited research exists with respect to the statistical attributes of the probability distribution of disturbances. Carey, Kwieciński (1994) firstly applied a shift exponential distribution and a uniform distribution to simulate the actual running time, which are composed of minimum running time and random disturbances. Carey, Carville (2000) used uniform distributions to model the running time extensions and beta distribution to generate the dwell time extensions. The exponential distribution is widely employed for generating entry delays, running time extensions, dwell time extensions and departure time extensions as disturbances to simulate the real operation in RailSys®. Furthermore, Cui et al. (2016) developed an algorithm inverting the corresponding parameters of disturbances distribution based on reinforcement learning in the framework of operational simulation. It solved the applicable problem more or less for different investigation areas with fine-tuned parameters.

2.2   Robust railway operation

In large amount of existing researches, robustness of railway operation becomes a key issue for handling the stochastic disturbances, so as to improve the resistance of the railway operation to the potential uncertainties.

However, there is still no consensus about the robustness definition, from the point of view of passengers or railway operators. On the one hand, Dewilde et al. (2011) describes robustness as the degree to minimize the passenger travel time in case of small disturbances; Takuchi et al. (2007) evaluates robustness using an index of passenger disutility with congestion rate, number of transfers and waiting time; Schöbel and Kratz (2009) defines robustness as the maximal size of the disturbances for which no passenger misses a transfer and how badly the passengers are affected by the disturbances. On the other hand,

Salido et al. (2008) points out that robustness refers to the ability to resist to imprecision; Goverde (2007) insists that a robust timetable should be able to compensate the delay of any train by time reserves, which keeps the delay from spreading over the entire network; Andersson et al. (2013) applies the term robustness as the timetable ability to handle small delays.

With discrepant definitions, various studies on robust railway operation are developed from both the planning stage (robust timetabling) and real time operation (robust dispatching) (see Figure 2-1).

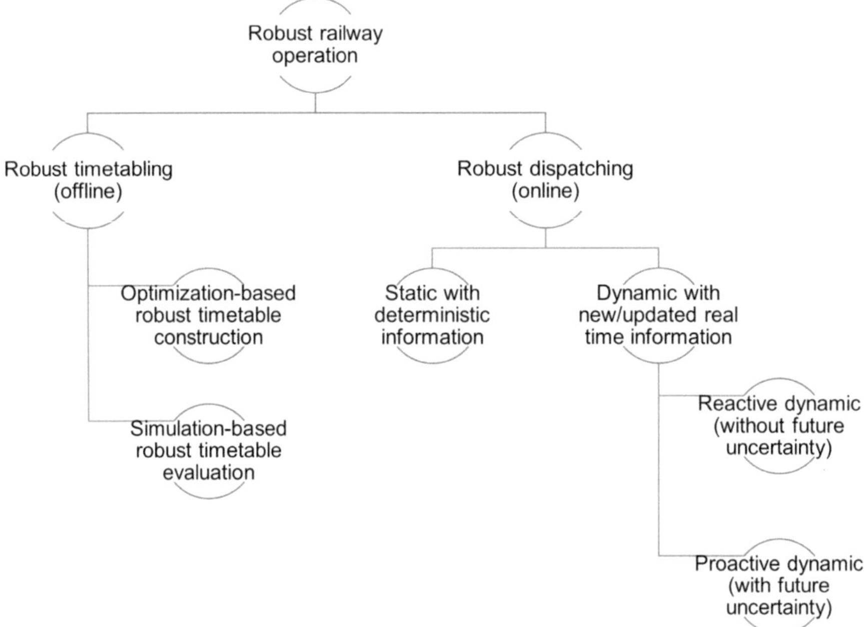

Figure 2-1   Review of literatures on robust railway operation

With the improvement of the computation tools, the dynamic dispatching is now widely conducted in the dispatching field and further developed as two major branches: reactive dynamic dispatching and proactive dynamic dispatching. They both get triggered by certain large deviations between the offline schedule and the real operation or the stages division points along the time dimension. The only difference between them is whether it takes future uncertainties into consideration when making decisions. In fact, the reactive dynamic dispatching (Lüthi 2009; Corman et al. 2011; Caimi et al. 2012), which neglects a view of future, is more prevailing because of its low computation complexity. Up to now, very limited studies focus on proactive dynamic dispatching, which considers perturbation probabilities and forecasts the future statuses. Lee, Ghosh (2001) developed a novel decentralized algorithm with soft reservation to schedule efficiently and mitigate the congestions in railway networks, which firstly emphasized the stability of solutions under different representative perturbations. Yet the generated perturbed traffic conditions are discussed separately and they refer to macroscopic, deterministic and approximated models. Meng, Zhou (2011) built a robust train dispatching model handling disruptions under the dynamic and stochastic environment. With a scenario-based solution search tree, the robust dispatching solution can be selected according to the expected delays with uncertain capacity breakdown duration (Duration of segment blocked is supposed to follow a Gaussian distribution). However, common dispatching strategies for handling disruptions such as re-platforming and rerouting are not considered and yet the closed loop control mechanism is left with only the conceptual illustrations. Quaglietta et al. (2013) further proposed a framework with several metrics to evaluate the stability of railway dispatching solutions, as time and knowledge of the perturbation progress. A typical Monte-Carlo scheme was applied for evaluating the chosen optimal dispatching solutions among the 30 randomly generated disturbed traffic scenarios (according to the probability distribution of entrance delays and dwell time extensions). It shows that it can provide suggestions for the designing the dispatching parameters (e.g. length of dispatching interval and prediction horizon) within the rolling time horizon framework, so as to obtain more stable dispatching solutions and minimize the number of dispatching instructions.

2.3 Objective systems of dispatching algorithms

An objective system serves as a benchmark to evaluate the quality of the dispatching solutions, which guide the dispatchers/dispatching software to generate dispatching solutions when conflicts or even worse deadlocks exist in railway operation. An inappropriate dispatching solution may lead to large consecutive delays, degrade customers' satisfaction, violate the realization of scheduled capacity, and in the worst case collapse the railway network. In (DB NETZ AG 2010), four major objectives are defined: 1) Restore the planed schedule in time in the operation control process; 2) Ensure the connection of consecutive trains; 3) Improve the overall punctuality (punctuality can be measured and evaluated by different indicators); 4) Maximize the utilization of the capacity of the railway network. The first objective emphasizes the speed of the schedule restoring, which not only testifies the quality of current dispatching solution but also saves lots of work for the dispatchers of the connected areas. The second one is primary to attract passengers and guarantee their satisfaction, which was considered in (Wegele et al. 2004; Larroche et al. 1996). As for the third one, since the perception of passengers related to punctuality is usually worse than reality, a high level of punctuality of train operations must be achieved in order to be attractive to passengers. The last objective is greatly related to the benefits of service providers. It is important to balance between extension of infrastructure, operating program and capacity of operation. A large capacity implies high efficiency of resources utilization, which should be maximized to obtain the profit maximization and adapt to the increasing travel demand by trains. In this project, the four objectives will be quantified and the comprehensive effect will be studied to select an optimal dispatching solution.

Among them, overall punctuality has been employed in majority of the existing researches regarding dispatching as a very significant indicator of service quality to the public. As for the reviewer's concern, punctuality can be perceived as trains' punctuality and passengers' punctuality. For instance, a train arriving at one station punctually doesn't mean that all passengers who take this train have arrived at their destination on time. In the situation that the connection of feeder trains fails, the trains' punctuality and the passengers' punctuality are not consistent. Yet somehow the passengers' punctuality highly depends on the trains' punctuality. Nowadays, several objective indicators regarding punctuality are widely used in the railway dispatching algorithms from both the passengers' and trains' aspects: minimization of total weighted waiting time (Hellström 1998; Lamma et al. 1997; Koch 2000; Li Ping et al. 2001; Törnquist, Persson 2007), minimization of total conveyance time/total trip time (Iyer, Ghosh 1995; Schlaich 2002), minimization of largest consecutive delay (Chiu et al. 2002; D'Ariano 2008) and minimization of passenger annoyance for platform changes and passenger waiting time (Wegele et al. 2004).

Objectives can be also changed with different scenarios in (Lüthi et al. 2007): the maximization of capacity is the objective when there are only small delays in the whole transportation system, while the assurance of the normal operation is the objective when infrastructure blockage occurs because of technical failures. Not only single objective oriented dispatching but also multi-objective oriented dispatching is developed. Punctuality and fluency of the traffic flow were taken as the dispatching objectives stepwise in (Martin 1995). Quality of transportation service with more emphasis on consumers was taken into consideration in (Larroche et al. 1996) based on an expert system. Setting of train routes, solving of train conflicts, and selection of station platforms were performed using a real-time software package to guarantee the capacity of infrastructure as well as the optimization of service quality.

2.4   Dispatching models

At present, various computer-based automatic/semi-automatic dispatching models have been proposed. According to the model definition and the algorithm used for resolving the solution, those models can be categorized into three types:

1) <u>Simulative model</u>. In simulative approaches, all train movements and operations are executed in a computer model which simulates the reality. The dispatching system is initialized with a basic timetable, and updated with real-time train movement information. Detecting potential conflicts and forecasting future train movements are performed iteratively. Once conflicts occur, a reschedule process based on a certain mechanism (synchronous or asynchronous) will be activated, and a new conflict-free timetable will be generated. Simulative methods can be classified as synchronous simulation and asynchronous simulation. The synchronous simulation is the most often used simulation model (Bidot 2003; Allan et al. 2004; Cheng 1998; D'Ariano 2008; Espinosa-Aranda, García-Ródenas 2012; Grube et al. 2011; Quaglietta et al. 2013) and is the basis of several software tools such as RailSys® and OpenTrack® (OpenTrack). The asynchronous simulation is depicted in the rescheduling chapter in (Hansen, Pachl 2014) and implemented in the software tool ASDIS (Asynchronous Dispatching).

2) <u>Analytical model</u>. In analytical approaches, train operations are abstracted to a mathematic optimization model with constraints. An objective function is firstly defined based on a specified optimization objective which reflects specific dispatching strategies. The optimization model is usually resolved based on a certain operations research algorithms, including linear programming (Bussieck et al. 1997; Boccia et al. 2013; Martin 1995; Törnquist, Persson 2007; Hutchison et al. 2009; Schlaich 2002), Branch-and-bound (Dorfman, Medanic 2004; D'Ariano et al. 2007; Higgins et al. 1997), Lagrangian relaxation (Keaton 1989; Brännlund et al. 1998) and so forth.

3) <u>Heuristic model</u>. Due to the computation burden and the complex nature of railway network, exact solutions of optimization models are hardly to be achieved in practice. Heuristic models balance the quality of the solution and the computational complexity. Within an iterative framework, heuristic model aims to find new solutions toward the direction of the optimal solution in each loop of iteration. A near-optimal solution will be derived by the end of pre-defined computation time or after certain number of iterations. In heuristic models, the evaluation method used to select better solutions can be based on a single objective (e.g. weighted delays or overall trip time) or a

combination of multiple criteria. In existing researches, various heuristic algorithms have been developed and employed, including genetic algorithm (Salim, Cai 1997; Wegele, Schneider 2004; Kacprzyk et al. 2008; Dündar, Şahin 2013), tabu search (Higgins et al. 1997; Gorman 1998; Törnquist, Persson 2005; Corman et al. 2010a), simulated annealing (Törnquist, Persson 2005; Sayarshad, Ghoseiri 2009) and problem space search (Albrecht et al. 2013; Pudney, Wardrop 2008), etc.

2.5   Dispatching assistance software

Based on the three mentioned dispatching models, different dispatching assistance software were developed, which helps the dispatcher to make decisions in order to avoid the improper solutions. In (Schaer et al. 2005), a flexible, modular and automatic dispatching assistant system in the Project DisKon was developed to solve conflicts in railway operation. The simulation software RailSys® provides a dispatching module to solve conflicts based on rules of thumb, and the influence on railway system capacity of corresponding dispatching parameters are discussed in (Martin et al. 2015). Decision Support Systems (DSSs) are developed for real-time railway traffic management by dividing the dispatching area into two subsections (centralized system and distributed system) in (Corman et al. 2010b). Overall performance is guaranteed by an overall feasibility test and repeated modification of local solutions, which solves dispatching problems quickly in a heavily used railway area. A computerized dispatching system based on global information with predicting potential conflicts, Railway traffic Optimization by Means of Alternative graphs (ROMA), is developed in (D'Ariano et al. 2008). Possible changes of train orders or routes can be suggested to the dispatchers and meanwhile the advisory speed profiles can be displayed to the train drivers. An approach for optimizing train driving pattern with consideration of dispatching, passenger information and recommendations is developed with software PULZURE in the project Zuglaufregelung (IEV 2005). MakSi-FM is an asynchronous automatic rescheduling system which contains algorithms to model infrastructure, timetable and construction works of rail tracks on a macroscopic level. Requirements such as resolution of occupation conflicts between trains, low computation time, network-wide application can be achieved with this system (Oetting 2013). As a subproject of FreeFloat, software KE/KL+ZLR was developed in the Project Regler by DB Netz AG, which provides automatic conflict detection and semi-automatic conflict resolution (Pänke, Klimmt 2012). Given that the knowledge of multi-scale simulations in railway planning and operation (Cui, Martin 2011), a multi-scale dispatching solution generator is developed in (Martin 2014b).

3 Operational risk analysis of block sections

3.1 Introduction to operational risk analysis

The railway system is a complex system with many interactive processes based on technical devices, human behavior, and the external environment. Therefore, it contains various risks for perturbations (Caimi et al. 2012). Occurrence of some disturbances can be prevented to some extent but unpredictable events are unavoidable and their consequences need to be analyzed, minimized, and communicated to the affected users (Törnquist 2012). Nowadays, inevitable operational disturbances are handled in two major ways simultaneously: robust train timetabling at the primary planning stage (Huisman et al. 2007; Kroon et al. 2008; Shafia et al. 2012; Andersson et al. 2013; Lindfeldt 2015) and synchronous dispatching in the case of occurrence of conflicts in real operation (Cheng 1998; Bidot 2003; Allan et al. 2004; D'Ariano 2008; Espinosa-Aranda, García-Ródenas 2012; Quaglietta et al. 2013).

Under all circumstances, identifying critical block sections, which are vulnerable to the operational disturbances and more likely lead to the collapse of railway operation, is the prerequisite for both of the two handling methods mentioned above. On the basis of block section vulnerability, it is capable of improving timetabling such as optimally setting limited time reserves (recovery time and buffer time) (Martin 2014a) on different block sections that vary with their vulnerability. By setting extra time reserves for vulnerable block sections and less time reserves for non-vulnerable block sections, primary delays as well as consecutive delays can be avoided without the degradation of railway network capacity. On the other hand, the vulnerability of block sections also plays a significant role for conducting more robust dispatching. In the phase of conflicts detections, the artificially prolonged blocking time on vulnerable block sections can provide a precise estimation in advance, so as to implement the dispatching solutions in time.

Studies on identifying the critical block sections in railway system are also known as the "bottleneck analysis". Up to now, there is no consensus on the definition of bottlenecks. A bottleneck is defined as "the decisive network element for the capacity performance, whose utilization rate lies in the deficient range of the quality" in (DB NETZ AG 2008). While by International Union of Railway *(UIC)*, a very highly utilized network element is identified as a bottleneck section of a railway network (UIC). In these two definitions, occupancy rate is the only criterion for identifying the bottlenecks. However, some infrastructure sections, though being highly utilized, have little influence on capacity performance. They might be misidentified as bottlenecks if only occupancy rate is taken into consideration. In view of this, Hantsch et al. (Hantsch et al. 2013) developed a new definition of bottlenecks as the infrastructure sections which may severely affect other train paths and the railway operation on adjacent sections, as well as leading to adverse effect on operating quality.

Among various definitions, different bottleneck analyses are conducted accordingly. Hartwig (2013) suggested that bottleneck analysis rely on the analysis of transport infrastructure, transportation model of demand prediction, and the average waiting time of existing timetable and conditions of the railway facilities. Drewello and Günther (2012) proposed bottleneck analysis beyond capacity discussion including the economical, spatial, and social context. Pöhle and Feil (2016) applied shadow prices of a developed column generation method as the indicator for identification of bottlenecks, which evaluated bottlenecks in a monetary unit. Li and Martin (2015) applied three indicators including bottleneck sensitivity, unfulfillable occupancy requirements, and occupancy rate to locate the bottleneck sections from operational perspectives. Besides, bottleneck significance and bottleneck relevance were firstly introduced to differentiate the bottleneck in the condition of a concrete traffic volume (bottleneck significance) and the potential bottleneck which appears when the traffic volume increases (bottleneck relevance). For both bottleneck significance and bottleneck relevance localization, a certain operating program is assumed, which means neither composition ratios of train types nor train sequences are changed in the timetable. Such bottleneck analysis results can be applied for capacity research of railway network but is contradictory in railway operation since dispatching changes the structure of the operating program often.

There is quite limited research on bottlenecks analysis based on the vulnerability of infrastructure sections. Andersson et al. (2013) discussed some disturbances sensitive locations as critical points in the process of quantitative robustness analysis and those points were identified simply through empirical observations of the Swedish timetable and traffic in 2011. Instead of critical points, the disturbances sensitive block sections are identified as bottlenecks in this study. The vulnerability of block sections, which refers to the susceptibility to disturbances that can result in considerable reductions in railway network serviceability, is reflected by the indicator "operational risk index". It is the expectation value of the negative impacts caused by the occurrence of disturbances on the specific block section. Different from the indicators (i.e., bottleneck sensitivity, unfulfillable occupancy requirements, occupancy rate) developed in (Martin, Li 2015) which focus on the hindrance on other trains operation caused by the occupancy of the infrastructure sections, the indicator "operational risk index" concentrates on the overall impacts caused by operational disturbances occurred on certain block sections. It provides a forewarning for the severe impacts possibly caused by disturbances.

Therefore, the results can serve as the foundation for developing an innovative dynamic dispatching algorithm with consideration of potential random disturbances. Furthermore, it can be also applied for effective timetabling at the primary stage, ranking the construction projects regarding to infrastructure expansion or maintenance and capacity research of the railway network. This study attempts to explore the negative impacts on the whole railway network caused by disturbances on each single specific block section instead of focusing on different utilization rate of each block section in real operation. On basis of the severity of the consequences and the provoked different operational risk levels, it is able to classify the block sections and conduct different dispatching or timetabling strategies accordingly. For example, let the self-recovery ability of the timetable to reduce the impact to a negligible level without dispatching actions when disturbances occurred on a low-risk block section, but reorder the trains sequence when a disturbance occurred on a high-risk block section.

3.2 Workflow of operational risk analysis

The risk R is defined as the product of the probable frequency of uncertain future events P and the expected loss of the corresponding events E (R=P·E) (Geiger 1994; Freund, Jones 2014). In this study, the operational risk of each block section is measured by the influence on the whole railway network when unexpected random disturbances occurred on that block section.

In the light of this idea, an effective approach to evaluate the risk level of a certain block section is to artificially impose random disturbances on that target block section and then simulate the railway operation based on the new disturbed timetable. The artificially imposed random disturbances are derived from the probability distribution of each type of operational disturbances based on Monte-Carlo simulation. Four types of random disturbances are generated on the target block section simultaneously for one disturbed scenario. Each timetable corresponding to a specific disturbed scenario are input to the available simulation tool RailSys®.

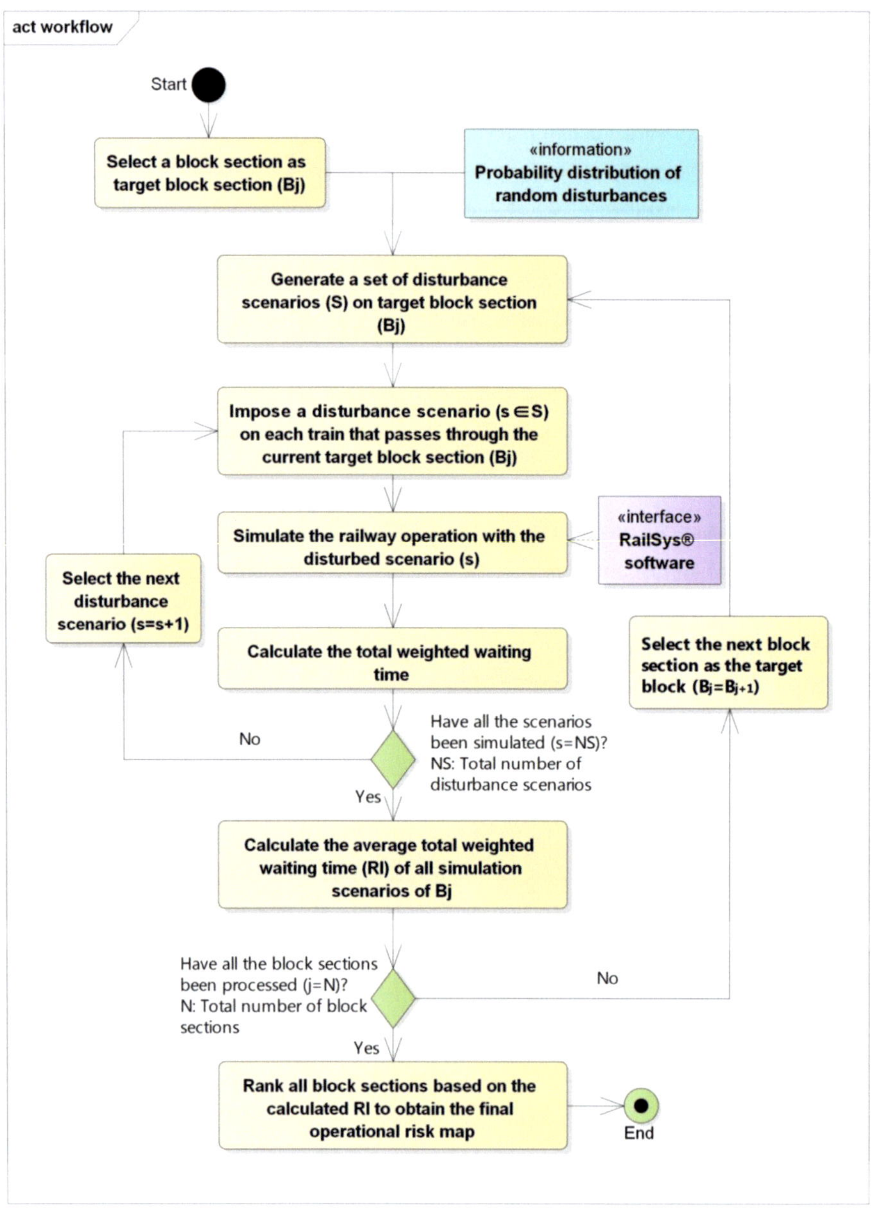

Figure 3-1   Workflow for mapping the operational risk

With the simulation results, the influence of the imposed disturbances on the whole network can be calculated by certain indicators such as overall punctuality, total weighted waiting time, delay coefficient, etc. Such process will be repeated multiple times to get the expectation value of the indicator, which will be utilized to measure the operational risk of block sections. After dealing with each block section by the way stated above, the operational risk map of the whole railway network can be obtained by comparing the value of indicators among all block sections.

Figure 3-1 describes the general flowchart of the proposed method, and details of each task will be illustrated above. In order to clearly demonstrate the iterative processes, the above stated process is summarized below by programming-like words.

For block_j=1; block_j ≤ number of block sections; block_j++

  For scenario_s=1; scenario_s ≤ number of simulation scenarios; scenario_s++

    Generate a disturbance scenario (disturbed timetable) for block_j;

    Simulate the disturbed timetable by railway simulation tool;

    Calculate the total weighted waiting time;

  Endfor

  Calculate the average total weighted waiting time of all scenarios for block_j;

Endfor

3.2.1 Determine the statistical distribution of random disturbances

As shown in Figure 3-1, the quantification of the random disturbances is the prerequisite for the subsequent simulations and analysis.

In theory, it is the best option to use the historical disturbances data collected from real world log files for this purpose. However, it is too difficult to obtain sufficiently such data from reality for the algorithm development and evaluation. Another alternative is to simulate the random disturbances based on certain statistical distributions.

In the current study, the main focus is to propose an algorithm framework for investigating the operational risk of railway block sections, and such framework will allow any distribution type of the imposed random disturbances. Researchers could define the distribution type based on the specific studied railway network or fit an empirical model if they have sufficient field data in hand.

In following, the proposed algorithm is illustrated by taking two typical types of distribution as examples: negative exponential distribution and Erlang distribution. Their probability distribution functions *(PDF)* are shown in Equation 3-1 (negative exponential distribution) and Equation 3-2 (Erlang distribution with the shape factor k as 2). For both distributions, the parameters β are the mean value of the disturbances.

$$f(x;\beta)=\frac{1}{\beta}e^{-x/\beta}, x>0 \qquad \text{Equation 3-1}$$

$$f(x;\beta)=\frac{4}{\beta^2}xe^{-2x/\beta}, x>0 \qquad \text{Equation 3-2}$$

In general, the stochastic disturbances are not always existed on every block section. Therefore, the train runs "with disturbances" and "without disturbances" need to be differentiated. For train runs "without disturbances", the generated disturbances are set as 0 on that block section, while for train runs "with disturbances", the generated disturbances are fit into the specific distribution.

With the PDFs described above, random samples of disturbances could be generated. For negative exponential model, a pseudo-random number ϑ between 0 and 1 was firstly produced from uniform distribution, and if $\vartheta <W_e/100$, a valid disturbance was calculated through Equation 3-3, otherwise the disturbance was assumed as 0. Here, the proportion parameter, W_e, controls the percentage of trains that might be influenced by random disturbances. In addition, to avoid that too large random samples were produced, D_{max} was used to define the upper limits of disturbance values. The parameters regarding to mean value of the disturbances (β), disturbances proportion (W_e) and maximum value of the disturbances (D_{max}) can simultaneously be calibrated based on the reinforcement learning calibration algorithm (Cui et al. 2016). For the Erlang distribution, two uniformly distributed random numbers, ϑ_1 and ϑ_2, were generated firstly, and the random disturbance was produced through Equation 3-4.

$$Dis = \begin{cases} \min\left(-\beta \cdot Ln\left(1-\frac{100v}{We}\right), D_{max}\right) & v < We/100 \\ \\ 0 & v \geq We/100 \end{cases} \qquad \text{Equation 3-3}$$

$$Dis = \begin{cases} \min\left(-\dfrac{\beta}{2} \cdot Ln\left(1-\dfrac{100v_1v_2}{We}\right), D_{max}\right) & v_1v_2 < We/100 \\ \\ 0 & v_1v_2 \geq We/100 \end{cases}$$

Equation 3-4

In this study, four types of disturbances will be taken into account:

1) <u>Entry delay</u>. It describes the difference between the scheduled arrival time and actual arrival time of involved trains at the boundary of the investigated area, which demonstrates the accumulated delays of the corresponding trains before they enter the investigated area.

2) <u>Running time extension</u>. It depicts the unscheduled extension of running time caused by the disturbances during train-running or the stochastic behavior of the train driver.

3) <u>Dwell time extension</u>. It represents the unscheduled time extension between the scheduled dwell time and the actual dwell time at each scheduled stop. For passenger train types, it is mainly caused by the time extension of passenger boarding and alighting. For freight train types, it is related to the unloading or loading time.

4) <u>Departure time extension</u>. It demonstrates the time extension after the process of passenger boarding and alighting or the freight loading/unloading is finished. It is caused by technical failures of the infrastructure ahead or the trains itself.

Among them, running time extension applies to all block sections, while entry delay is specific for the block section which is the entrance of the investigated railway network, departure time extension and dwell time extension only applies for those block sections which contains a train station or a scheduled stop.

3.2.2 Generate and impose the disturbances on target block section

For each train that passes through a certain block section (denoted as B_j, $1 \leq j \leq$ the number of block sections in the railway network), a set of random samples of each type of disturbance that may occur on that block section is generated by sampling in a standard Monte-Carlo scheme based on their PDF obtained in Section 3.2.2 (following negative exponential or Erlang distribution). From the basic timetable, scheduled arrival and departure time of each train operating on block section B_j are known, and then the generated disturbance samples are imposed on those trains when they operate on B_j. Consequently, a disturbed timetable is obtained, and it is defined as one "disturbance scenario" of B_j.

For the target block section B_j, three attributes in the basic timetable might be changed, including departure time, running time, and dwell time (specific for trains with stop). In detail, the running time (Run_{dis}) and dwell time ($Stop_{dis}$) in the disturbed timetable are the sum of their original values in basic timetable (Run_{basic}, $Stop_{basic}$) and the produced random samples of running time extension (ΔRun) and dwell time extension (ΔDw), respectively (Equation 3-5, Equation 3-6); while the disturbed departure time (Dep_{dis}) is its basic time (Dep_{basic}) postponed by the sum of entry delay (ΔEn), departure time extension (ΔDep), dwell time extension (ΔDw) and running time extension (ΔRun) (Equation 3-7).

$$Run_{dis} = Run_{basic} + \left(\Delta Run \right)$$

Equation 3-5

$$Stop_{dis} = Stop_{basic} + \left(\Delta Dw \right)$$

Equation 3-6

$$Dep_{dis} = Dep_{basic} + \left(\Delta En + \Delta Dep + \Delta Dw + \Delta Run \right)$$

Equation 3-7

Table 3-1   An example of disturbance scenario of a certain block section

Train ID	Basic timetable		Random disturbances (s)				Disturbed timetable	
	Departure time point	Running time (s)	Entry delay (s)	Departure time extension (s)	Running time extension (s)	Dwell time extension (s)	Departure time point	Running time (s)
T1	00:15:20	50	91	0	58	0	00:16:51	108
T2	00:21:00	32	125	0	20	0	00:23:05	52
T3	02:05:41	50	0	0	20	0	02:07:17	70
T4	03:25:05	32	0	0	23	0	03:27:13	55
T5	04:59:17	32	0	0	32	0	05:00:07	64
T6	05:07:02	50	134	0	48	0	05:09:16	98

Table 3-1 gives an example of the disturbance scenario: the basic timetable for B_j is listed in the left two columns. For disturbances, no dwell time extension and departure time extension is imposed since B_j doesn't contain a regular train stop; entry delays are only occurred on

Train T1, T2, and T6; while running time extensions influence all trains that pass through B_j. As results, the disturbed timetable is calculated and listed in the right two columns.

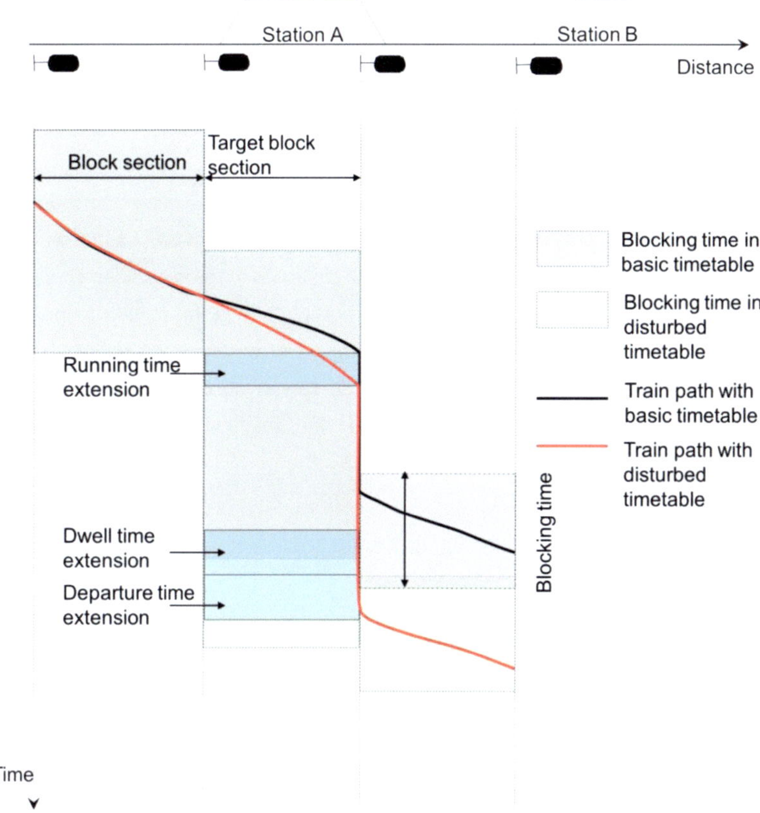

Figure 3-2   Blocking time diagram of a single train on block sections in part of its train path with basic timetable (black) and disturbed timetable (red), respectively.

To illustrate the influence of the imposed random disturbances, Figure 3-2 displays one part of a certain train path of a single train with its basic timetable and disturbed timetable, respectively. It is seen that the second block section is the target block section on which disturbances are imposed. As results, at station A, the train is delayed by disturbances with dwell time extension and departure time extension; and during running on the target block section, the train is delayed by disturbance with running time extension.

3.2.3 Calculate the indicator of operation risk analysis

With the generated disturbed timetable, railway operation simulation was performed with the aid of RailSys® software. Based on the simulation results, it is possible to measure the influence of the imposed disturbances on the whole railway network, and further to evaluate the operational risk of the target block section. For this purpose, in this study, the total weighted waiting time (*TotalwWT*, Equation 3-8) is taken as the indicator of the block section operational risk.

The actual occupation of each train on every block section can be calculated based on the data records in simulation protocol. By comparing the actual occupation (simulation of the disturbed timetable) of the related train on every block section with the scheduled occupation (simulation of the basic timetable), the unscheduled waiting time of the related train can be calculated. Here, in order to quantify the different impact of unscheduled waiting time for different train types, the total weighted waiting time, served as expected loss for each random disturbances scenario, is represented as the weighted sum of each train's unscheduled waiting time.

$$TotalwWT = \sum_{t=1}^{M}\left\{ w_t \sum_{j=1}^{N}\left[\left(e_{t,j} - b_{t,j}\right) - \left(\bar{e}_{t,j} - \bar{b}_{t,j}\right)\right]\right\} \qquad \text{Equation 3-8}$$

Where

TotalwWT	total weighted waiting time
t	Train index
j	BlockID
M	Total number of trains
N	Total number of block sections on a train path
w_t	Weight of train t for calculating the total weighted waiting time
$\bar{e}_{t,j}$	End of blocking time of train t on block section j in the disturbed timetable
$e_{t,j}$	End of blocking time of train t on block section j in the simulation results
$\bar{b}_{t,j}$	Start of blocking time of train t on block section j in the disturbed timetable
$b_{t,j}$	Start of blocking time of train t on block section j in the simulation results

The weight system applied for the calculation is determined based on the reduction rate per additional minute of delay, which is stated in the train path pricing system (DB NETZ AG

2015). The logic behind this weight system is the different perception of customers with regard to delays of different train types. For instance, customers who paid more money for express trains deserve better operation performance with sufficient punctuality. Therefore, under the circumstances of delays, a compensation system regarding to different train types for every additional minute of delay is enacted by Deutshce Bahn. This compensate system is represented by the reduction rate per additional minute of delay, reflecting the perceptual negative influence regarding to different train types to some degree. For long-distance passenger transport, the reduction rate is 3 times more than that of freight transport, while for regional passenger transport, it is 2 times as many as that of freight transport. Accordingly, the weight for different train types is set as 3 (long-distance passenger transport); 2 (regional passenger transport) and 1 (freight transport).

3.2.4   Obtain the operational risk map

In the framework of operational risk analysis, only one single disturbance scenario is far from enough to represent the expectation value of the indicator, *TotalwWT*. To ensure the statistical reliability of matching the disturbance distributions, multiple disturbance scenarios are generated and are further simulated in RailSys®. The calculated *TotalwWT* of all the scenarios are averaged to be taken as the risk index (*RI,* Equation 3-9) of the target block section.

In addition, to reveal the relative relationship of operational risk among all block sections rather than the absolute value of *RI*, a normalization procedure can be applied to calculate the normalized *RI* (*NRI*, Equation 3-10) among all block sections after the *RI*s for all block sections were obtained. Then, the procedures stated above will be implemented on each block section in the investigated railway network respectively. With that, all block sections could be ranked and classified to a certain number (*L*) of levels based on the calculated *RI* to obtain an operational risk map for the railway network.

$$RI = \frac{1}{NS}\sum_{s=1}^{NS} TotalwWT_s \qquad\qquad \text{Equation 3-9}$$

$$NRI_j = \frac{RI_j - RI_B}{RI_{max} - RI_B} \qquad\qquad \text{Equation 3-10}$$

Where

RI Risk index

NRI_j	Normalized risk index of block section j
s	Scenario index
NS	Number of disturbance scenarios
RI_B	Averaged RI of basic timetable
RI_{max}	Maximum RI value among all block sections

3.3 Sensitivity analysis of parameters on operational risk

In addition, to investigate if, and to what degree, the operational risk of block sections has dependence on the basic timetable or the disturbances distribution, the procedures stated above are also employed to different basic timetables and with disturbance samples being generated from different distribution types. The simulation results will be analyzed and compared with respect to:

1) Different distribution types of imposed disturbance;

2) Different traffic volumes in the investigation area;

3) Different freight transport shares in basic timetables.

To be clear, some timetable related terms used in this section are stated as follow:

1) Basic timetable: the original timetable without any modification;

2) Disturbed timetable: generated by imposing disturbances on the basic timetable;

3) Condensed basic timetable: generated by compressing the basic timetable to different proportions for compact schedules without changing the mixture of train types (to keep the structure of the operating program);

4) Condensed and disturbed timetable: generated by imposing disturbances on the condensed basic timetable.

3.3.1 Influence of distribution types of imposed disturbances

Robustness of the operational risk analysis result to different distribution types of imposed disturbances is significant to ensure the applicability of the proposed approach, since there is no such a standard distribution type for stochastic disturbances applicable everywhere. In fact, the distribution of operational disturbances and its corresponding parameters differs from place to place, varies over time and highly depends on train routes. Therefore, different distribution types for generating the imposed disturbances should be considered and the comparison between the resulting operational risk maps is necessary.

3.3.2 Influence of traffic volume in the investigation area

To investigate how the traffic volume influences the *RI* and *NRI* of each block section, the basic timetable was compressed by different proportions and the procedures stated above were implemented on those condensed timetable respectively. The higher the proportion is, the more condensed the timetable is, and more trains are operated in the railway network. In a congested railway network, the disturbances may lead to more severe negative influences, triggering high operational risk. The condensed timetables are generated automatically with the application of the software PULEIV (Martin et al. 2008) (Programm zur Untersuchung des Leistungsverhaltens), further developed by Martin and Chu (Martin, Chu 2014) with randomness and fixed mixture of train types, taken as the condensed basic timetable for conducting the operational risk analysis.

3.3.3 Influence of different freight transport schedules

Basically, for a certain railway network, the basic schedules for long distance passenger transport (FRz) and regional passenger transport (NRz) are relatively stable and will be implemented for a long time once in operation; while freight transport (NGz) schedule is relatively flexible and is prone to change due to the fluctuated goods transport needs and the congestion degree of the whole network. Accordingly, a test was conducted by investigating how the simulation results change if the freight transport schedules vary.

3.4 Determination of influencing parameters

3.4.1 Number of operational risk levels

In addition, the value of risk level L_{total} (see Section 0) is supposed to be defined in advance. It can be manually set or automatically derived based on certain criteria.

In theory, if multiple groups of the aforementioned process are performed by setting different value of L_{total} to each group, the larger the L_{total} is, the more likely that the same block section will be classified in different operational risk levels in different groups, and vice versa. Therefore, the most appropriate L_{total} should make the variance of block section risk levels over multiple times of experiments as small as necessary, and meanwhile lead the classification resolution to be as fine as necessary.

In order to identify the most appropriate L_{total} to achieve the balance, a trial-and-error test was performed by setting L_{total} from 1 to N_{block} (the number of block sections in the studied railway network) in turn. For a specific L_{total}, the risk level classification results of all block sections under each condensed basic timetable could be obtained.

Then, three calculation procedures were taken:

1) *Cal 1*: For every block section, calculate the standard deviation (*Std*) of its risk levels among all condensed basic timetable (Equation 3-11);

$$Std = \sqrt{\frac{\sum_{i=1}^{N_{cond}} \left(L_i - \bar{L}_{cond} \right)^2}{N_{cond}}} \qquad \text{Equation 3-11}$$

2) *Cal 2*: For every block section, calculate the normalized standard deviation (*NStd*) based on *Cal* 1 (Equation 3-12);

$$NStd = \frac{Std}{\bar{L}_{cond}} \qquad \text{Equation 3-12}$$

3) *Cal 3*: Calculated the average value of the *NStd* (Averaged *Nstd*) among all block sections (Equation 3-13).

$$Averaged \ \ Nstd = \frac{\sum_{j=1}^{N_{Block}} NStd_j}{N_{Block}} \qquad \text{Equation 3-13}$$

Where

N_{cond} Number of condensed basic timetable

N_{block} Number of block sections

L_i Risk level of a certain block section under the *i-th* condensed basic timetable

$\bar{L}_{cond}$ Mean value of L_i among all condensed basic timetables

In this manner, a relationship between the *Averaged Nstd* and L_{total} can be drawn. Usually the *Averaged Nstd* increases monotonically with L_{total}. An appropriate L_{total} can be chosen according to the degree of increase afterwards (see Section 0).

3.4.2   Number of disturbance scenarios

Another issue worth mentioning is the number of simulations, that is, the disturbance scenarios (NS), for each block section. In theory, a too small NS is not enough to ensure the statistical reliability and extreme results might be generated; while a too large NS will increase the computation burden and be very time-consuming. To achieve the balance and decide the most appropriate NS, a test was conducted by increasing NS from 1 with increment of 1, and for each NS, the RIs of all scenarios are calculated for each risk level of block sections respectively. When RIs becomes stable and shows no variation for all the curves of each risk level as the NS increases, the appropriate NS can be selected. The parameter NS can be different from case to case and vary with different size of investigation area, therefore, the test must be conducted in advance to save the computation efforts, and meanwhile to achieve convincing results.

4   Dispatching algorithm development in rolling time horizon framework

In this chapter, a hybrid dispatching algorithm is proposed based on both heuristic and simulation approaches within a rolling time horizon framework (Peeta, Mahmassani 1995).

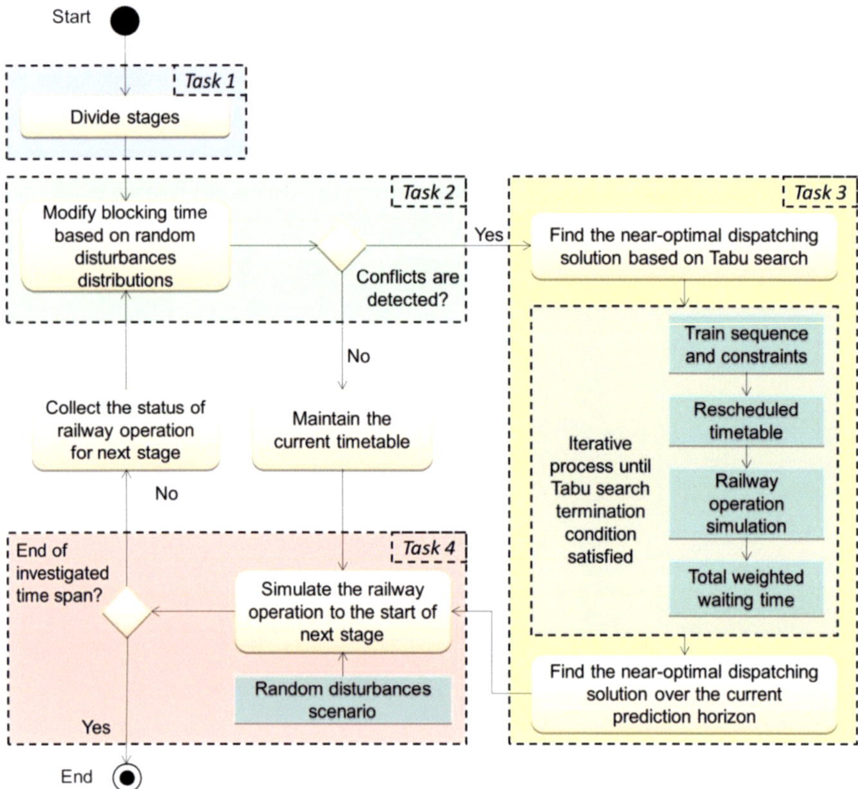

Figure 4-1   General flowchart of the dispatching algorithm development

The framework divides the dynamic (i.e. with information and status known only as time goes by) dispatching problem into multiple static (i.e. deterministic and with complete information) sub-stages that can be solved by static dispatching methods. For each sub-stage, potential conflict detection is firstly performed by taking operational risk oriented disturbances into account. Given that conflicts detected, a heuristic approach, Tabu search in particular, is employed to search for the near-optimal dispatching solution. Based on the generated dispatching solution, a railway simulation tool, as the proxy of real-world railway operation during the algorithm development, is used to simulate the revolution of the railway system till the start of next sub-stage. The procedures described above will be repeated for all the stages till the end of the investigated time span. A general flowchart of the dispatching algorithm is shown in Figure 4-1, and details are described in the four separate procedures.

4.1 Stage division

The basic idea of the rolling time horizon framework is to segment a dynamic process into multiple static sub-processes and handle each sub-process separately. To this end, the stage division is the prerequisite of the entire algorithm framework.

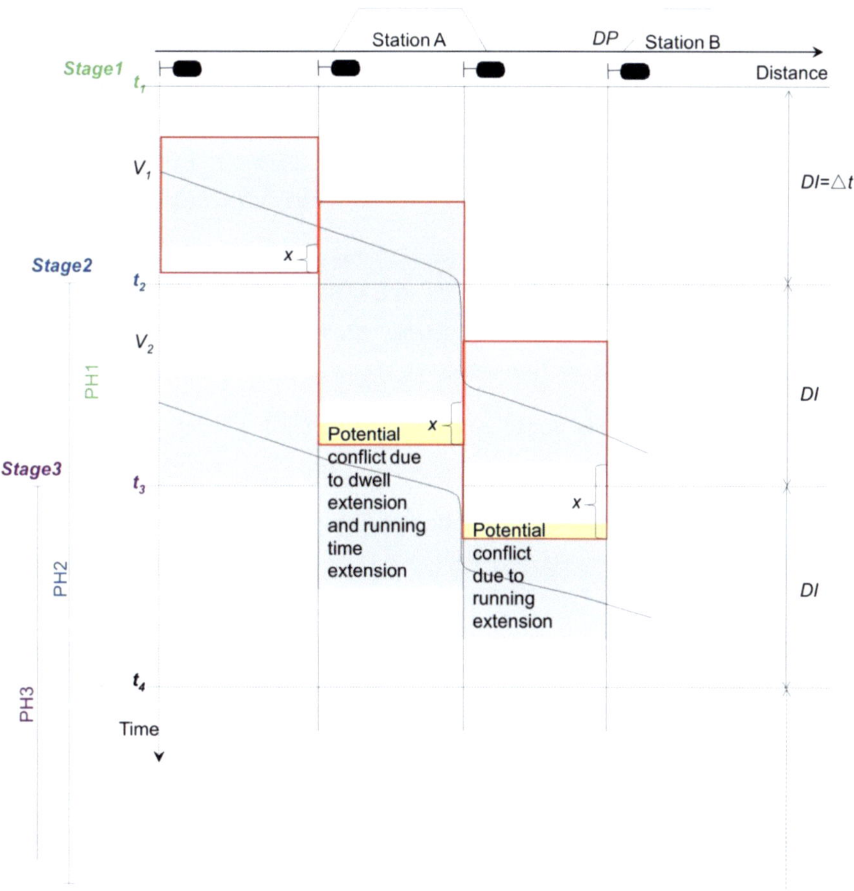

Figure 4-2   Rolling time horizon framework for the proposed dispatching algorithm

Figure 4-2 represents a conceptual diagram of rolling time horizon approach with two consecutive trains (V1, V2). For the distance dimension (horizontal direction), the train paths are divided into multiple track sections named "block section". For the time dimension (in vertical direction), the entire investigated time span is decomposed into multiple time periods named "stages" which are partially overlapping and spaced at fixed time intervals named "dispatching interval" (DI, as value of Δt). The time length of each stage is identical and denoted as prediction horizon (PH) which is usually an integral multiplier of DI (i.e., $PH = K^*DI$, $K \in N+$).

Within this framework:

1) At the beginning of each stage (t_1, t_2, t_3...), potential conflicts detection is performed within the time period of corresponding PHs ($PH1$, $PH2$, $PH3$...). According to the detection results, dispatching solutions are generated including reordering, retiming, or combination of the both. In the case of no potential conflicts detected, no dispatching is needed and the original schedule will be maintained.

2) From the beginning of each stage (t_1, t_2, t_3...), the generated dispatching solutions are conducted till up to the beginning of the next stage, during which the time span is the DI. For example, the rescheduled timetable based on the dispatching solutions made at time point t_2 will be operated between t_2 and t_3.

3) The rolling procedures stated above will be repeated until the end of the whole investigated time span.

4.2 Risk-oriented conflicts detection at a single stage

Potential conflicts detection is performed at the beginning of a certain stage (t_1, t_2, or t_3...). This task quantitatively identifies possible conflicts that will occur in future PH period on certain block sections. According to the blocking time theory proposed by (Hansen, Pachl 2014), the blocking time of a block section for a train without a scheduled stop contains time for clearing the signal, time for the driver to view the approach indication, the approach time between the approach indication signal and the entrance signal, the running time between the block signals, the clearing time and the release time. Among them, the running time between the block signals consists of the pure technical running time (train operation with the maximum speed) as well as the recovery time. An occupancy conflict can be detected when an overlap between the blocking times (grey area in Figure 4-2) of two trains is observed on the same block section. Note that in term of dispatching, the recovery time, which is included in the blocking time, can be applied to compensate the occurred disturbances. During the process of rescheduled timetable generation, if the sum of all kinds of disturbances is smaller than the recovery time, the timetable will be kept as the original one. In this case, only changing the speed profiles of the train operation is sufficient.

Different from existing research that predict the railway states only based on deterministic conveyance times (i.e. running times and/or dwell times based on the scheduled timetables), this study detects occupancy conflicts based on modified blocking time under the consideration of operational risk-oriented disturbances in order to take account of random disturbances that occur normally and frequently in real-world railway operation.

For this purpose, artificial disturbances are imposed on all block sections before checking if overlaps exist between the blocking time of two (or more) trains on a certain block section. Here, four types of disturbance mentioned in Section 3.2.2 are considered. For different block sections, the amount of the imposed disturbances is different, and such differences correspond to the operational risk levels of block sections that have been obtained in Chapter 3. In other words, on the basis of the results of Chapter 3, a higher risk block section is subject to be imposed more disturbances when detecting potential conflicts, and vice versa. The underlying reason is that a higher risk block section (assumed as B_h) will have larger impact on the whole railway network once disturbances occur on B_h, and thus, B_h should be treated more seriously than relatively lower risk block sections when deciding if potential conflicts will probably exist on B_h.

Accordingly, it is necessary to establish a functional relationship between the amount of imposed disturbance (x) and the risk level (L) of a block section:

$$x \rightarrow f(L) \qquad\qquad \text{Equation 4-1}$$

However, it is difficult to directly link x to L through any known quantitative models, and thus an intermediate variable is necessary to build the relationship between x and L. Considering that x follows a certain statistical distribution, an alternative way is to use the cumulative distribution function (CDF, denoted as P) of that distribution as a bridge to connect x and L:

$$x \rightarrow P \rightarrow f(L) \qquad\qquad \text{Equation 4-2}$$

For the left part ($x \rightarrow P$), it is easy to link x to P by the integration of the probability distribution function (PDF, denoted as $f(x)$) that x follows:

$$P = F(x) = \int_{-\infty}^{x} f(x)dx$$

Equation 4-3

By inversing the above equation, x can be described as a function of P:

$$x = F^{-1}(P)$$

Equation 4-4

For the right part ($P \rightarrow f(L)$), in this study, P is assumed to be linearly related to L with maximum-minimum normalization. It should be noted that other linear models or non-linear models may also be appropriate to use only if the model gives a monotonic positive relation between P and L.

$$P = \frac{P_{max} - P_{min}}{L_{max} - L_{min}}(L - L_{min}) + P_{min}, \quad L_{min} \leq L \leq L_{max}, \quad L \in N$$

Equation 4-5

By combining Equation 4-4 and Equation 4-5, x can be described as a function of L in the form of:

$$x = F^{-1}\left(\frac{P_{max} - P_{min}}{L_{max} - L_{min}}(L - L_{min}) + P_{min}\right), \quad L_{min} \leq L \leq L_{max}, \quad L \in N$$

Equation 4-6

Where F^{-1} is the inverse function of the *CDF* that x follows; P_{max} and P_{min} are pre-defined parameters within the rage of [0, 1]; L_{max}, L_{min}, and the independent variable L of specific block section have been obtained in Chapter 3. Based on Equation 4-6, it is obvious that the amount of imposed disturbances for a block section is determined by the operation risk level of that block section which is the output of Chapter 3.

Figure 4-3 illustrates a concrete example of the deductions stated above. Assuming that the running time extension (x) for a certain railway network follows the negative exponential distribution with the rate parameter β = 10 minutes. The *PDF* of x can be expressed as Equation 4-7 and plotted in Figure 4-3a:

$$f(x) = \frac{1}{\beta} \cdot e^{-x/\beta} = \frac{1}{10} e^{-x/10}, \quad x \geq 0$$

Equation 4-7

By the integration of *PDF*, the *CDF* of x can be obtained as Equation 4-8 and plotted in Figure 4-3b:

     Hybrid Model for Proactive Dispatching of Railway Operation under the Consideration of Random Disturbances in Dynamic Circumstances

$$P = F(x) = \int_0^x \left(\frac{1}{10} \cdot e^{-x/10} \right) dx = 1 - e^{-x/10}, \quad x \geq 0 \qquad \text{Equation 4-8}$$

It can be seen that a larger x corresponds to a larger cumulative probability (P), which means that the larger the given x is, the higher the possibility that a random sample of disturbance is smaller than x. Then, the inverse function can be obtained as Equation 4-9 and plotted in Figure 4-3c:

$$x = F^{-1}(P) = -10 \cdot Ln(1 - P) \qquad \text{Equation 4-9}$$

On the other hand, let P_{max} = 0.95, P_{min} = 0.05, L_{max} = 5, and L_{min} = 1 for example, the relationship between P and L can be specified as Equation 4-10 and plotted in Figure 4-3d:

$$P = 0.225L - 0.175, \quad 1 \leq L \leq 5, \quad L \in N \qquad \text{Equation 4-10}$$

By combining Equation 4-9 and Equation 4-10, the functional relationship between x and L can be specified as Equation 4-11 and plotted in Figure 4-3e:

$$x = -10 \cdot Ln[1 - (0.225L - 0.175)], \quad 1 \leq L \leq 5, \quad L \in N \qquad \text{Equation 4-11}$$

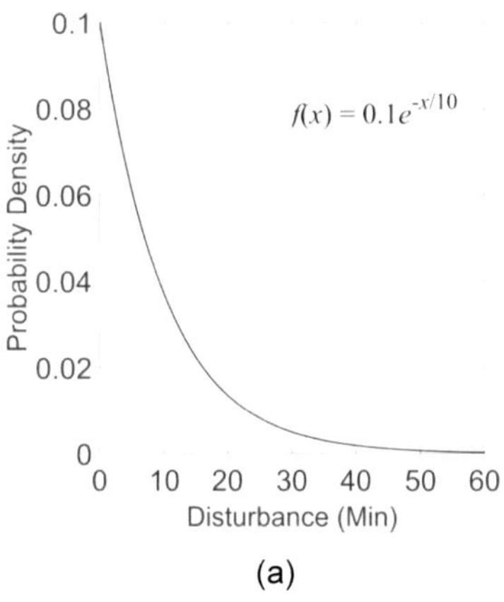

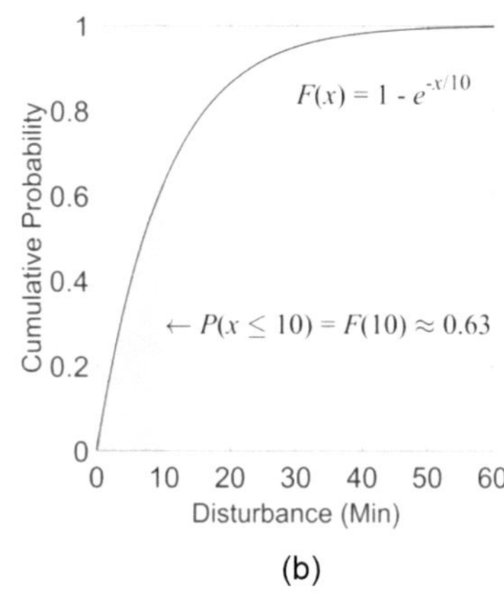

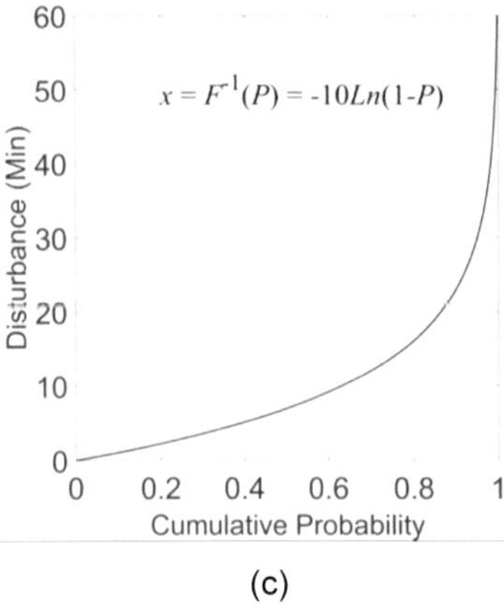

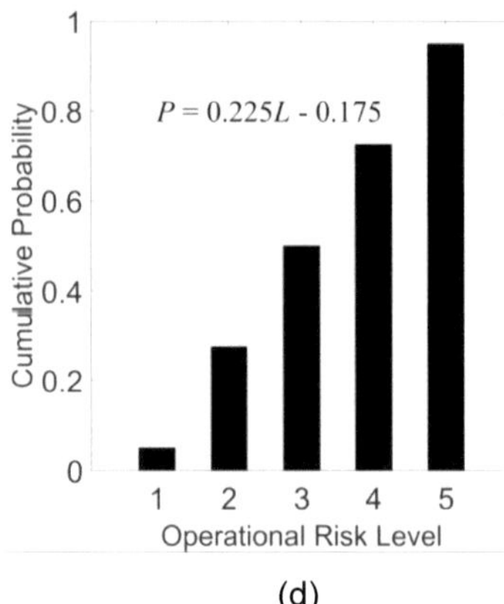

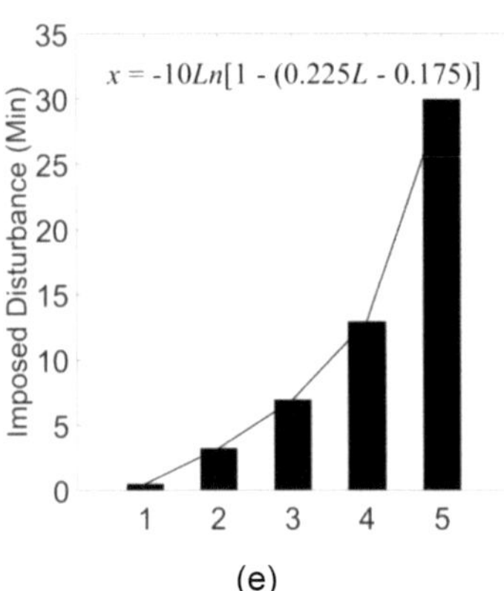

Figure 4-3   a) Probability distribution function *(PDF)* and b) cumulative distribution function *(CDF)* of running time extension with β = 10 min; c) inverse function of *CDF*; d) linear relationship between operational risk level *(L)* and the cumulative probability *(P)* that the imposed disturbance should corresponds to (with P_{max} = 0.95, P_{min} = 0.05, L_{max} = 5, and L_{min} = 1); e) functional relationship between the imposed disturbance and the operational risk level of block sections.

With potential disturbances imposed, taking Figure 4-2 for example, modified blocking time of train V1 is then obtained (the red box). The conflict is identified if the red box bottom line of V1 overlaps with the current scheduled blocking time area of train V2 (yellow area). Similarly, the conflict between V2 and V3 (not shown in Figure 4-2) will be detected by comparing the modified blocking time of V2 with the current scheduled blocking time of V3 over each block section. This procedure is repeated within *PH*1 over all trains on their train paths to complete the conflicts detection at stage 1.

4.3   Dispatching solutions at a single stage

With the conflicts detection results obtained at the beginning of a certain stage, a dispatching solution for that single stage can be calculated. This dispatching solution includes retiming and reordering (Figure 4-4).

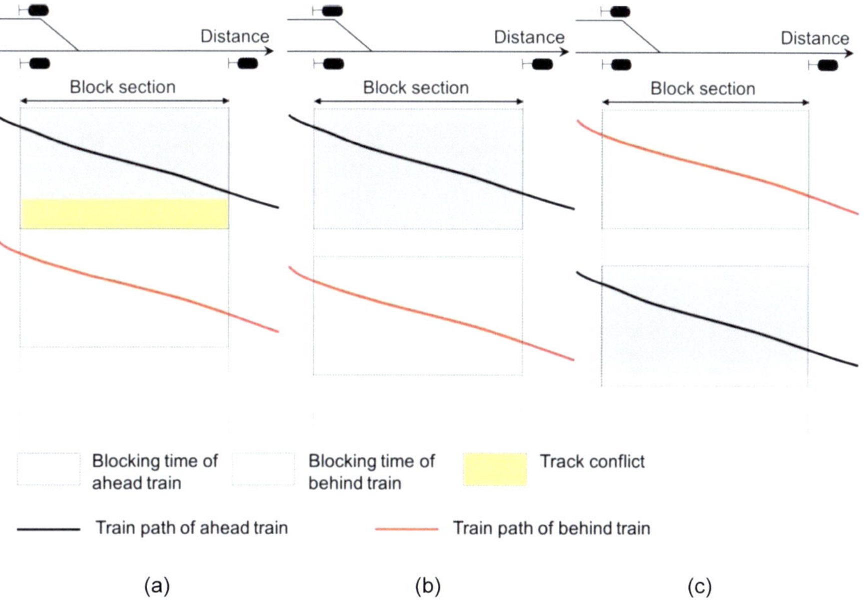

Figure 4-4   a) Basic schedule, b) retimed schedule, and c) reordered schedule of two consecutive trains on a certion block section.

Retiming only alters the arrival/departure time of the trains in conflict, while reordering changes the train sequence passing certain dispatching point (*DP*) through loops or junctions, at which overtaking or crossing can only take place at loops if the track is long enough.

In this study, Tabu search (TS) algorithm (Glover 1986), one of the typical heuristic approaches, is employed to find the near-optimal train orders. Combined with appropriate retiming which is used to solve slight conflicts, the rescheduled timetable is generated and operated for the current stage. Figure 4-5 summarizes the flowchart of dispatching procedures that are performed within a single stage defined in Section 4.1.

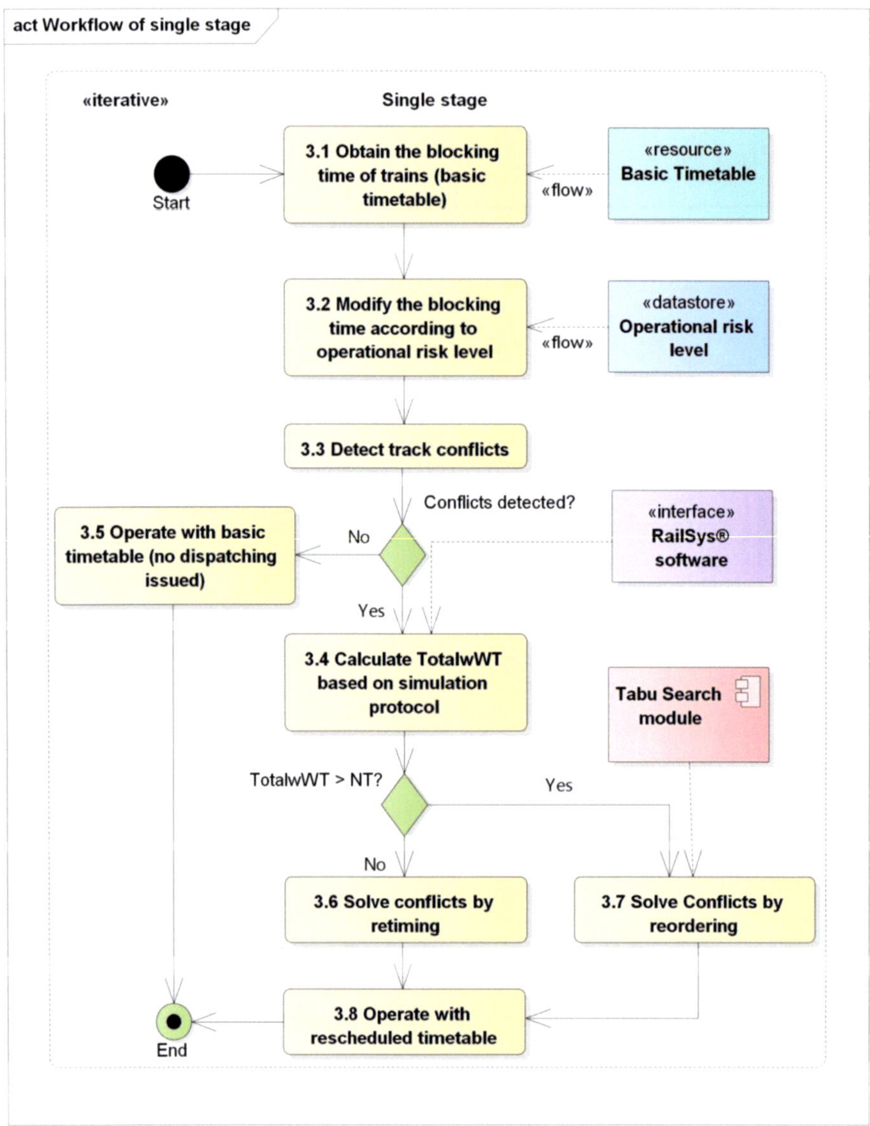

Figure 4-5   Dispatching procedures performed within a single stage

4.3.1   Initial simulation of the disturbed timetable

Given that potential track conflicts are detected, RailSys® software is utilized to simulate the railway operation with the *PH* period based on the disturbed basic timetable, which is generated by artificially imposing risk-oriented disturbances on the basic timetable as stated in Section 4.2. Similar to the risk analysis in Chapter 3, the total weighted waiting time ($TotalwWT$) is calculated based on the simulation results. If $TotalwWT$ is smaller than a pre-defined threshold NT, the detected conflicts in the current stage will be treated as slight conflicts, otherwise they will be considered as significant conflicts. For the former case, simple retiming is enough to solve, while for the latter case, reordering is firstly considered based on the Tabu search algorithm until $TotalwWT$ reduced to NT or smaller.

4.3.2 Retiming for slight conflicts

As shown in Figure 4-5, slight conflicts are to be handled by retiming (Figure 4-4b) which simply postpone the time schedule of the trains that at relative subsequent sequence to avoid the overlap of blocking time on the same block section, without changing the order of trains passing that block section.

4.3.3 Reordering for significant conflicts using TS algorithm

Tabu Search algorithm (TS), created and formalized by Glover (Glover 1986, 1989, 1990), is a metaheuristic search method employing local search methods used for mathematical optimization. Since the proposal of the TS algorithm, it has been widely used in various researches especially in the field of transportation (Gorman 1998; Törnquist, Persson 2005; Corman et al. 2010a; Zufferey, Verma 2011; Zhang et al. 2012; Chen, Niu 2013; Sterzik, Kopfer 2013).

TS uses a local or neighborhood search procedure to iteratively move from one potential solution (S) to an improved solution (S') in the neighborhood of S, until some stopping criterion has been satisfied (generally, an attempt limit or a score threshold). Local search procedures often become stuck in poor-scoring areas or areas where scores plateau. In order to avoid these pitfalls, TS algorithm is capable of avoiding such a dilemma by constraining a potential solution from returning to recently visited areas of the search space, referred to as cycling. The strategy of the approach is actually to maintain a short term memory of the specific changes of recent moves within the search space and preventing future moves from undoing those changes. This memory structures form that is known as the Tabu list, a set of rules and banned solutions used to filter which solutions will be admitted to the neighborhood to be explored by the search. In its simplest form, a Tabu list is a short-term set of the solutions that have been visited in the recent past (less than n iterations ago, where n is the number of previous solutions to be stored). More commonly, a Tabu list consists of solutions that have changed by the process of moving from one solution to another.

Figure 4-6 displays how the TS algorithm is employed in this study to solve the track conflicts by searching for a better train sequences passing the conflicted block section. Following the flowchart shown in Figure 4-6, technical details and key procedures of the TS algorithm will be described below:

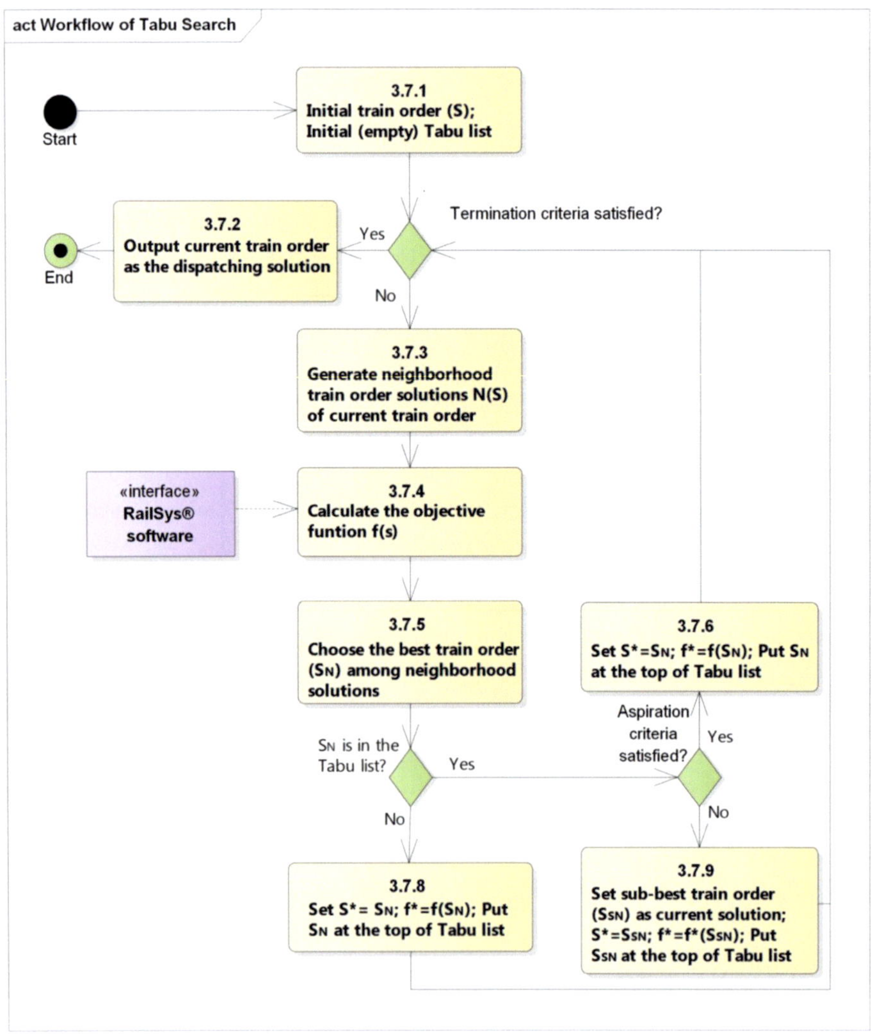

Figure 4-6   Technical flowchart of Tabu search algorithm for finding the near-optimal train orders as dispatching solutions

1) <u>TS-1, Initialization:</u> Train orders are initialized as that in the current timetable which will potentially induce track conflicts. Tabu list, with a pre-defined length (L_{TL}), is initialized as an empty list. In this study, the element in Tabu list is a train sequence passing a certain block section.

2) <u>TS-2, Neighboring solution:</u> Generate all immediate neighbors of the initial train sequence as candidates of solution by neighboring move. An immediate neighboring move, as depicted in Figure 4-7, is defined as the switch of the orders of two trains in a train sequence passing a certain block section.

Figure 4-7   A sample of the immediate neighboring move of the train sequence. a) Original train sequence; b) Neighboring move by switching the train order of V2 and V4

3) <u>TS-3, Objective function $f(S)$:</u> Simulate the railway operation on the basis of each candidate solution using RailSys® software. For each simulation result, calculate the value of the objective function. In this study, the total weighted waiting time ($TotalwWT$), which was defined in Equation 3-8 in Chapter 3, is taken as the objective function. Among all candidate solutions, the one that corresponds to the minimum $f(S)$ of the simulation result is selected as the improved solution (S').

4) <u>TS-4, Tabu list:</u> Check if S' has already been in the Tabu list. If not, set S' as the incumbent best solution ($S^* = S'$), set $f(S')$ as the incumbent best value of the objective function $f^*(S)$, and meanwhile, put S' at the top of the Tabu list; If so, move to TS-5.

5) <u>TS-5, Aspiration criteria:</u> Aspiration criteria are employed to override a solution's Tabu state. Provided that a solution is "good enough" according to a measure of quality or diversity, it will be treated as S^* even if it has been included in the Tabu list. A simple and commonly used aspiration criterion is to allow solutions which are better than the currently-known best solution. Otherwise, the secondary best solution is taken as the incumbent best solution and is put at the top of Tabu list.

6) <u>TS-6, Termination criteria:</u> The user-specified termination criteria are set to stop the search procedure. Two examples of such conditions are a simple search loop limit (N_{loop}) or a threshold on the objective function value. Before the termination criteria are met, the

Tabu search loop will continue searching for a more optimal solution, which will repeat from TS-2.

4.4 Iterative railway simulation and dynamic dispatching in the rolling time horizon framework

Following the procedures stated in Section 4.3, the near-optimal dispatching solution of a certain single stage is obtained by the combination of reordering and retiming, and the rescheduled timetable is generated accordingly. Then, the railway operation will stick to such rescheduled timetable until the next dispatching time point (i.e., the beginning of the next stage) of which the time interval is DI (see in Figure 4-2). It should be noted that at the algorithm development stage, the RailSys® software is used as the proxy of real-world railway operation to simulate the railway revolution based on the rescheduled timetable. To be closer to reality, a random disturbance scenario is imposed on all trains in each block section during the simulation. Such scenario is generated by sampling disturbances in a standard Monte-Carlo scheme based on their probability distribution functions.

However, a dispatching solution generated for a certain stage might be degraded due to random disturbances or unexpected events. In this study, therefore, a dynamic dispatching process is realized in the rolling time framework. Figure 4-8 shows the conceptual diagram of such a process. In detail, potential conflicts detection and dispatching are conducted iteratively at the beginning of each stage. If no conflicts identified, the current timetable (that might be the basic timetable if no dispatching actions have been executed before the current stage) is still operated without any modification. Otherwise, the procedure described in Section 4.3 is conducted again to update the near-optimal dispatching solution and is operated/simulated for the corresponding stage. This iterative process will not be terminated until the end of the entire investigated time span.

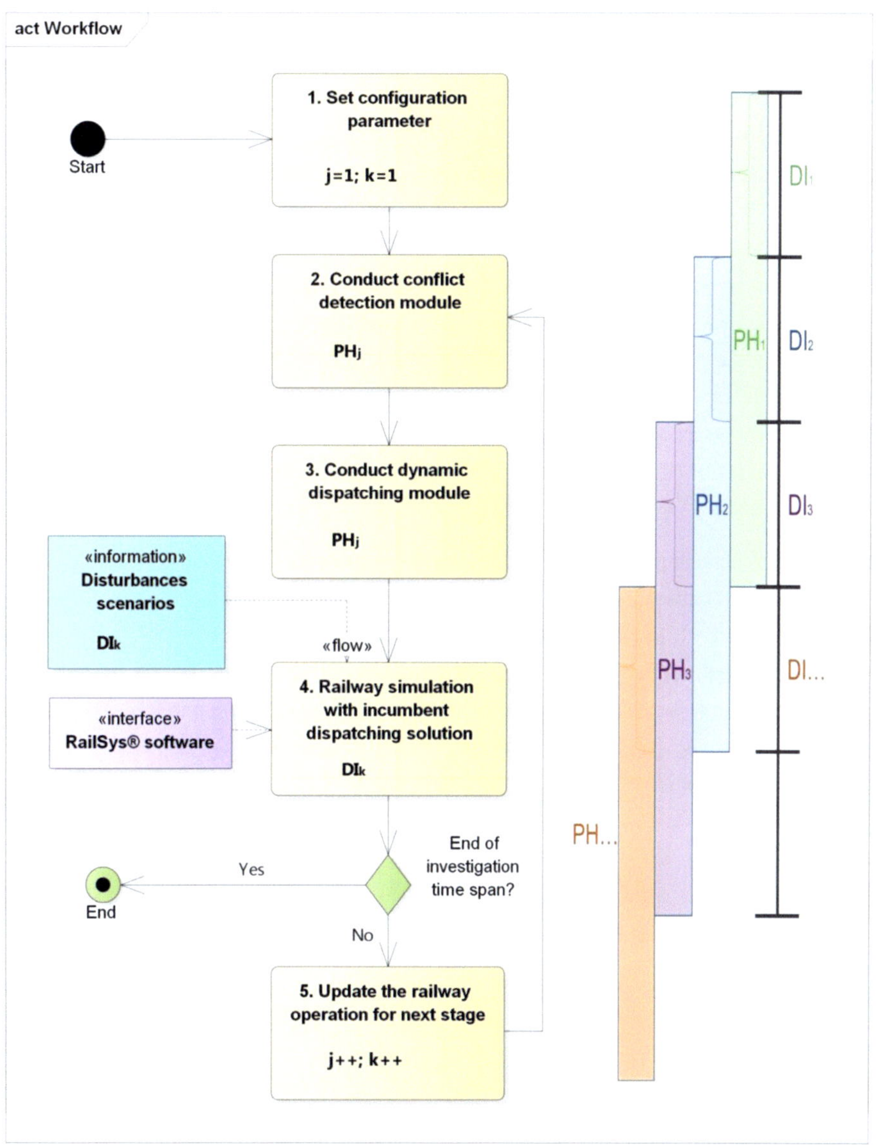

Figure 4-8   Conceptual diagram of dynamic dispatching process in the rolling time horizon framework

Special attention should be paid on the time overlaps between different stages, which suggests that although conflicts detection and dispatching solution are made for the future *PH* period, it is operated/simulated only for future *DI* period that is shorter than *PH*. In this way, it is allowed that *PH* (i.e., the considered time range) can be long enough to obtain reliable conflicts detection results and robust dispatching solutions; while *DI* can be short enough to deal with unexpected disturbances and generate new dispatching solutions responding to the change of railway states. In such a rolling dynamic framework, it is able to collect the information of the actual railway operation regularly and update the dispatching solution in a timely manner. Also, for the same reason (*PH* > *DI*), it basically takes a certain time span for the dispatching solutions to be in effect, meaning that the obtained dispatching solution for future *PH* period may not include any action in the immediate following *DI* period. In this situation, the current timetable will be stuck to till the beginning of the next stage, and then, the results of conflicts detection and resolution for the next stage will supersede the dispatching solution of the current stage.

For a concrete example, the dispatching solution generated at 1:00pm is designed by minimizing the total weighted waiting time of the whole railway network between 1:00pm to 2:00pm (here, PH=1 hour). At 1:20pm (DI=20 minutes), railway operation status is checked again. If potential conflicts are detected, a new dispatching solution will be designed by minimizing the total weighted waiting time between 1:20pm to 2:20pm and take place of the previous dispatching solution. If no further conflict is detected, the dispatching solution generated at 1:00pm will be still in effective. Then, at 1:40pm, 2:00pm, 2:20pm......, the above stated procedures will be repeated until the end of the investigation time span (e.g., 00:00am).

5 Assessment of the proposed dispatching algorithm

In Chapter 3 and Chapter 4, an operational risk oriented dispatching algorithm was proposed in a dynamic framework with the consideration of potential random disturbances. This chapter will describe how to assess the performance of the proposed algorithm. At first, three indicators will be employed to evaluate the effectiveness and robustness of the dispatching solution. Based on those indicators, inter-comparisons will be conducted between basic timetable and rescheduled timetable in which the dispatching solution is implemented. In addition, an existing dispatching method, FCFS, will be compared with the proposed algorithm in this study to demonstrate the advantage of the latter.

5.1 Indicators for assessing the performance of the dispatching solution

5.1.1 Total weighted waiting time

Total weighted waiting time (*TotalwWT*, Equation 3-8) was the primary metric of operational risk of block sections in Chapter 3, as well as the objective function in the Tabu search procedures for searching for the optimal/near-optimal dispatching solutions in Chapter 4. It is widely used to indicate the reliability and punctuality of scheduled train services, which is highly useful for railway infrastructure managers and train operators in planning, management and marketing of these services, and for customers in making travel choices (Yuan 2006). To be an effective dispatching solution, there should be a reduced *TotalwWT* after the dispatching action is implemented. Otherwise, the dispatching fails to improve the railway operation. Thus, the *TotalwWT* is taken as the foremost indicator to evaluate the quality of dispatching solution generated at each stage, as well as the overall influence on the railway network in the whole investigation time span.

5.1.2 Number of relative reordering

The number of relative reordering (*NRR*) measures the total number of trains that have been reordered with respect to the timetable scheduled at the immediate previous stage (Quaglietta et al. 2013). *NRR* is one of the metrics that represent the robustness of the dispatching solution. A larger *NRR* means more frequent reordering of train sequences and brings to the railway network more burden and uncertainty; while a smaller *NRR* suggests a more stable railway operation. In general, a train is defined as reordered if it has a sequence number larger than its predecessors. Table 5-1 shows an example for three consecutive stages the lists of train order passing a certain block section, as well as the calculation of *NRR*.

Table 5-1 Lists of train sequences passing a certain block section with respect to the timetable for three consecutive stages and correspoding *NRR*.

Stage	Train sequence	Sequence number at current stage (j)	Sequence number at last stage (j-1)	NRR
1	T1	1	1	0
	T2	2	2	
	T3	3	3	
	T4	4	4	
2	T2	1	2	1
	T1	2	1	
	T3	3	3	
	T4	4	4	
3	T4	1	4	2
	T3	2	3	
	T2	3	1	
	T1	4	2	

If the sequence number attributed to trains (T1, T2, T3, and T4) follows the same order of the dispatching solution at stage 1, the *NRR* to the dispatching solution of stage 2 is 1, since train T1 has a larger sequence number at stage 1 than that of stage 2, and T1 is therefore reordered with respect to the solution at stage 1. Similarly, if the sequence number follows the order given by the dispatching solution at stage 2, the amount of reordering relative to the timetable of stage 3 is 2 because both trains T1 and T2 are reordered.

5.1.3 Average absolute retiming

Average absolute retiming (*AAR*) measures the time shift between the train paths provided by a rescheduled timetable and those of the basic timetable (Quaglietta et al. 2013). Quantitatively, at a generic time *t*, *AAR* can be calculated as Equation 5-1:

$$AAR(t) = \frac{1}{M} \sum_{j=1}^{M} \left| t - \tau_{j,loc}^{TT}(t) \right|$$

Equation 5-1

Where

M total number of trains running at time *t*

loc	location reached by train j at time t according to the rescheduled timetable
$\tau_{j,loc}^{TT}(t)$	passing time of train j at location *loc* as in the basic timetable: for stations with a stop, this is the time published in the timetable; on block sections, this is the planned time at which the train is expected to pass, following an unhindered path, and considering running time supplements.

5.2 Comparison with FCFS dispatching principle

In the circumstance of occupancy conflicts, several possible dispatching decisions can be made to solve the conflicting situations: First-Come First-Serve (FCFS) principle, First-Come Last-Serve (FCLS) principle, cancel train runs, shorten the train path and operate on alternative routes. Among them, FCFS is the simplest and usually applied "default" dispatching principles in many simulation tools. Within the framework of FCFS principle, the conflicting train, which requests the infrastructure resources earlier, possesses the higher priority. Even though the FCFS principle is not always the optimal choice for solving the occupancy conflicts, it is widely used for the comparison and evaluation with other dispatching approaches.

Similarly in this study, the dispatching solutions generated by FCFS and the proposed dynamic dispatching algorithm are also evaluated through the three indicators described in Section 5.1, respectively. With such an inter-comparison, the advantages of the proposed algorithm can be manifested.

5.3 Comparison with real time dispatching decisions

Nowadays, the dispatching process is still conducted manually in general. It is usually supported by assistant software, e.g. LeiDis-S/K (Leitsystem Disposition Strecke/Knoten) in DB Netz AG's operation control center, which provides the dispatchers with actual operation data. Based on the differences between the actual states (positions/time) of each train and the scheduled ones, the resolution of conflicts is provided by the dispatchers considering certain rules of thumb (see Section 1.1). However, due to the inherent complexity of dispatching tasks, experienced dispatchers not always dedicate the same dispatching solutions even under the same circumstance. For example, for the dispatchers with experience on both field dispatching and train monitoring on assistant software, the dispatching solutions include not only the conventional retiming and reordering at an early stage, but also unconventional solutions such as double overhauls, re-platforming; while for the other dispatchers, unconventional solutions are seldom carried out in practical.

In order to testify the practicability of the proposed dispatching algorithm and promote the algorithm into implementation, a comparison between the solutions generated with the proposed algorithm and the solutions came up with the field dispatchers is necessary when the dispatching protocols in the real operation is available. The three indicators described in Section 5.1 can be employed to conduct the comparisons. Deviations between the solutions can be recorded for further analysis, including:

1) Whether the prediction of the proposed conflict detection is correct or overestimated.

2) Whether the deviations exist in the same or similar conflict situation;

3) In which circumstance the unconventional dispatching solutions (not included in the proposed algorithm) can achieve better performance.

In this manner, the proposed dispatching algorithm can be further improved and adjusted.

6 Sensitivity analysis of dispatching parameters

In the dispatching algorithm proposed in Chapter 4, several user-defined parameters are important to obtain high-quality dispatching solutions. Table 6-1 lists those parameters and summarizes how they have influences on the dispatching process in general. In this chapter, quantitative sensitivity analysis on these parameters will be introduced based on the three indicators described in Section 5.1.

Table 6-1   Dispatching-related paramters and their general influence on the dispaching process

Parameter	Denotation	Too large	Too small
PH	Prediction length of each stage in the rolling time horizon framework (Figure 4-2)	High complexity and computation burden	Dispatching solutions lose efficacy at relatively far future or deadlock problems are neglected
DI	Dispatching interval in the rolling time horizon framework (Figure 4-2)	Fails to predict and solve potential conflicts in time	Redundant dispatching calculation
P_{max}	The parameter which controls how much disturbance imposed on the highest operational risk block section (Equation 4-6)	Cause significant delay and degrade railway capacity	Fails to predict potential conflicts
N_{loop}	The upper limit of TS search loops used as the TS termination condition (Section 4.3.2)	Redundant search and waste of time	Quality of the dispatching solution can't be ensured

6.1   Sensitivity analysis of parameters in rolling time horizon framework

As listed in Table 6-1, PH and DI are the two key parameters in the rolling time horizon framework. PH is the time span of a single stage which determines how long the conflicts detection and dispatching look forward. In theory, a longer PH includes more trains into consideration which increases complexity and computation burden during the dispatching process. However, a too short PH is prone to generate a dispatching solution that only handles local conflicts but lose efficacy to solve potential coming conflicts due to the unexpected disturbances. Thus, a balance is need to be achieved by selecting a most appropriate PH. DI controls the frequency at which conflicts detection is performed and dispatching solution is updated. A shorter DI will probably cause redundant dispatching calculation; while in the case of a longer DI, the railway operation will stick to one timetable too long and fails to predict and solve potential conflicts in time. Similar to PH, an appropriate DI is also important to ensure the balance stated.

Considering that PH and DI are both time related parameters and the dispatching process is influenced by the combination of them rather than any individual one, the sensitivity analysis of these two parameters are performed concurrently. For this purpose, PH is set as different proportions (10%, 20%, 30%, 40%, 50%, 60%, 70%, 80%, 90%, 100%) of the entire investigated time span; as DI should be shorter than PH base on the definition of rolling time framework and PH is usually an integrally multiplier of DI, DI is set as different proportion (5%, 10%, 20%, 25%, 50%) of each set PH. For all combinations of PH and DI, the dispatching algorithms are tested and the $TotalwWT$, NRR, AAR are calculated and compared to quantify the influences of the two parameters and select the optimal combination.

6.2   Sensitivity analysis of parameters in random disturbances generator

According to Equation 4-6, in the process of conflicts detection, P_{max} and P_{min} control how many random disturbances are imposed on block sections with the highest and lowest operational risk level, respectively. Based on the operational risk analysis described in Chapter 3, block sections with the lowest risk are likely to have ignorable impacts on the entire railway network even if relatively severe disturbances occur on them. Therefore, it is reasonable to fix P_{min} as a very small value (e.g., 0.05) and in this way, the amount of disturbances imposed on block sections with other risk levels are totally determined by P_{max}. Here, P_{max} is set to be varying from 0.30 to 0.95 with interval of 0.05 to test the results of the dispatching algorithm. For each P_{max}, the $TotalwWT$ is calculated and compared. Considering that a too large P_{max} may cause significant overall delay of the entire railway network, the time of the last operating train in the network is also extracted and compared.

6.3   Convergence analysis of the TS parameter N_{loop}

The number of search loops is taken as the termination condition in the TS algorithm for search the optimal or near-optimal dispatching solution. Since TS algorithm is a local optimization method, N_{loop} has a direct influence on the quality of the final output solution. Considering that significant disturbances will probably involve more trains into conflicts and thus more candidate solutions exist in the process of Tabu search, the convergence analysis of N_{loop} should be related to P_{max} stated above. In each case of P_{max}, the convergence point of N_{loop} are to be found and compared with P_{max} to see how P_{max} has effects on the optimization of dispatching solutions.

As stated in Section 4.4, random samples of disturbances are imposed to the railway network during the RailSys® simulation to reflect the real-world situation as close as necessary. Consequently, a single test of the sensitivity analysis of those dispatching related parameters is not enough to report the results with high confidence. Therefore, extensive tests are needed to ensure the statistical reliability of the parameter analysis.

7 Case study

In this chapter, a case study will be conducted to illustrate how to use the proposed algorithm and assess its performance. Section 7.1 introduce the investigated railway network; Section 7.2 gives results of the operational analysis stated in Chapter 3; Section 7.3 analyzes results of the dispatching algorithm proposed in Chapter 4; Section 7.4 evaluates the proposed methods using several metrics mentioned in Chapter 5; Section 7.5 performs calibration of some dispatching related parameters; and Section 7.6 conducts a comparison between the proposed algorithm and other dispatching methods.

7.1 Investigation area

The proposed dispatching algorithm deals with an entire railway network, not only a single train run or line respectively. A classical small network for algorithm development and evaluation shown in Figure 7-1 is taken for the case study in this dissertation. It consists of 35 block sections including four train stations (AHX; BS; EN; LBC). Three types of trains, including long distance passenger transport (FRz), regional passenger transport (NRz) and freight transport (NGz), are operated in the system. With different operating programs, specifically traffic volume (trains/hour) and mixture of train types (% of train types), the opposing, merging, following and crossing conflicts can be included and studied in the scope of this investigation area, which ensures the developed dispatching algorithm is capable of handling various types of conflicts and enhances the extensibility to large railway networks. The investigation time span considered in this dissertation is 6 hours.

It is worth mentioning that according to specific research requirements, the railway infrastructure can be described in different models:

1) Macroscopic model (abstract node-link model)

   In a macroscopic model, the node stands for a station/junction, while the link stands for the open track between stations/junctions. It is able to cover large investigation area. However, simplification of significant areas may fail to detect certain conflicts.

2) Mesoscopic model (station track groups-route nodes-infrastructure element lines model (Nießen 2008))

   In a mesoscopic model, the node is further divided into station track groups and route nodes due to the different functionality. In this way, the occupancy and utilization of nodes can be investigated more in detail by application of queue theory. However, due to the lack of precise platform information, the deviations of the calculated waiting time from reality can be hardly avoided.

3) Microscopic model (basic structure (Martin et al. 2012), component of train paths)

   Both the macroscopic and mesoscopic model are used only for a rough evaluation of a railway network. For the purpose of a detailed conflict detection and resolution, a microscopic infrastructure model is needed. In this study, a "two-level infrastructure model (Martin, Li 2015)", which consists of basic structure and component of train paths, can be adopted.

   Basic structure is the maximum occupation unit allowed to be occupied by only one train simultaneously on a microscopic level without direction information. The boundaries of a basic structure can be the closest signal, signal release point, route release point, or the borderline of the investigation area. It can be generated automatically from any track layout of a network. Considering that it is divided by the operational points in bi-direction, it is unnecessarily small which may lead to huge computation time in the process of dispatching.

   Component of train paths is the smallest directional occupancy elements on a train path which can be divided. A single component of train path can be occupied by one train at most simultaneously. It is applicable for both the fixed block system and the moving block system. In a fixed block system, every train path can be demonstrated as a series of them. Under the circumstances of partial route releasing is not considered, the component of train paths can be regarded as the block section, defined as a section of track in a fixed block system which a train may enter only when it is not occupied by other vehicles in (Hansen, Pachl 2014). It facilitates the calculation of the blocking time or conflict detection within acceptable computation time. Therefore, the block section is considered in this study as the befitting infrastructure model. Yet in this manner, further consideration of partial route releasing can be also investigated.

The proposed algorithm framework is not limited to block sections but also compatible with other infrastructure models. Figure 7-1 displays an overview of the studied railway network with the BlockID defined in this work. Within it, each block section is taken as the basic study unit.

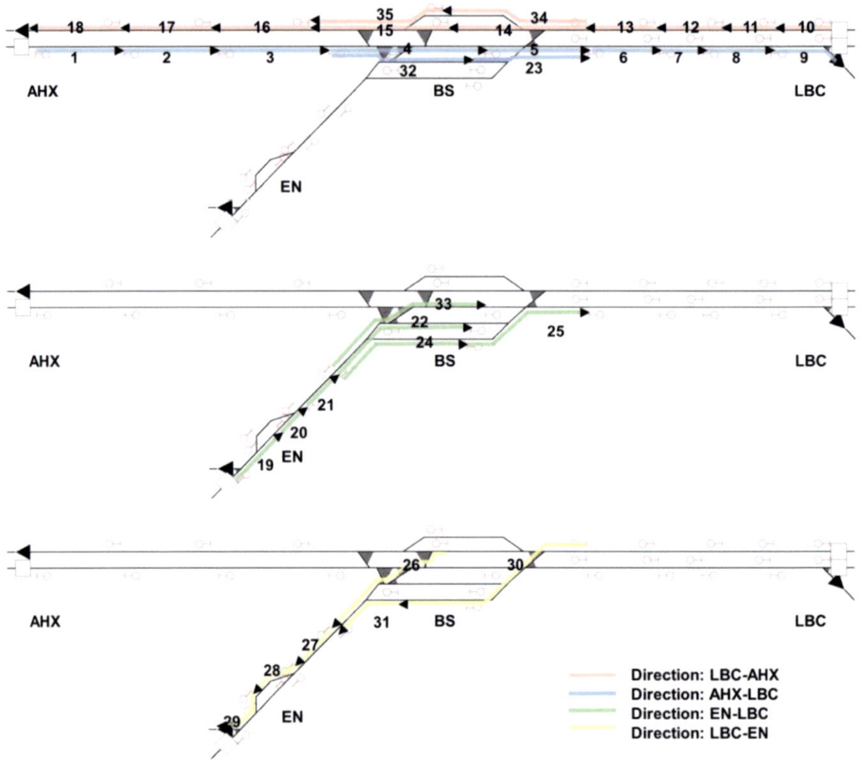

Figure 7-1   The railway network for case study

7.2   Operational risk analysis

In this section, the operational risk analysis stated in Chapter 3 is tested with the investigated railway network described in Section 7.1. For each disturbance scenario of a target block section, the corresponding disturbed timetable was fed in RailSys® software to simulate the railway operation. A certain number of scenarios were generated and simulated for each block section, which then enabled the *RI* for that block section to be calculated through Equation 3-9, and the corresponding *NRI* to be calculated through Equation 3-10. In this section, the simulation results will be analyzed and compared.

7.2.1   Distributions and parameters of random disturbances

As mentioned in Section 3.2.1, two distribution types of random disturbances are employed in this study. Table 7-1 provides the parameters considered for each type of disturbance and for each train class in this study. To be comparable, the same parameters are used for both of them to generate the random samples.

Table 7-1   Parameters of the four types of disturbance for different train types

	FRz			NRz			NGz		
	β (min)	D_{max} (min)	We (%)	β (min)	D_{max} (min)	We (%)	β (min)	D_{max} (min)	We (%)
Entry delay	5	60	50	2	30	50	20	60	50
Departure time extension	1	5	5	1	5	5	5	15	10
Running time extension	1	30	40	1	15	60	10	30	60
Dwell time extension	1	5	5	0.5	5	5	5	15	10

Considering the entry delay for example, Figure 7-2 displays the PDFs for the three train classes based on Equation 3-1 and Equation 3-2, respectively.

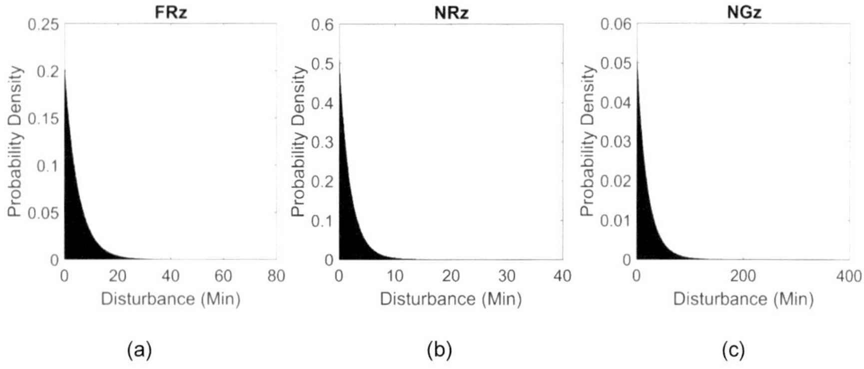

(a)   (b)   (c)

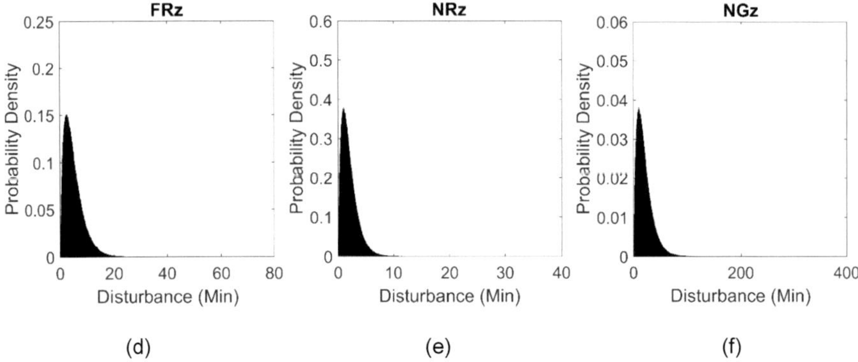

(d)   (e)   (f)

Figure 7-2 The probability distribution of entry delay for train types of FRz, NRz, and NGz following a), b), c) negative exponential distribution and d), e), f) Erlang distribution, respectively

In this case:

1) A similar trend is displayed for both distributions with respect to their right tails, while the negative exponential distribution presents more potential to produce small disturbance values than the Erlang distribution;

2) The largest mean value for all train classes is that of entry delay, followed in a decreasing manner by running time extension. Departure time extension or dwell time extension is the smallest value in comparison;

3) NGz is more likely to have larger mean value than FRz and NRz for all cases of disturbance due to the lower priority possessed by freight trains, not including express freight transport service.

A significant number of disturbed timetables were generated through the use of the PDFs mentioned above, along with their corresponding parameters (see Section 3.2.2). In Figure 7-3, a disturbance scenario for block section 8 (B8) is displayed within the railway network shown in Figure 7-1. Within it, a global view of the influence on all trains which pass through the certain target block section B8 is presented, showing the difference between basic and disturbed timetables of all trains operating on B8 in different time periods.

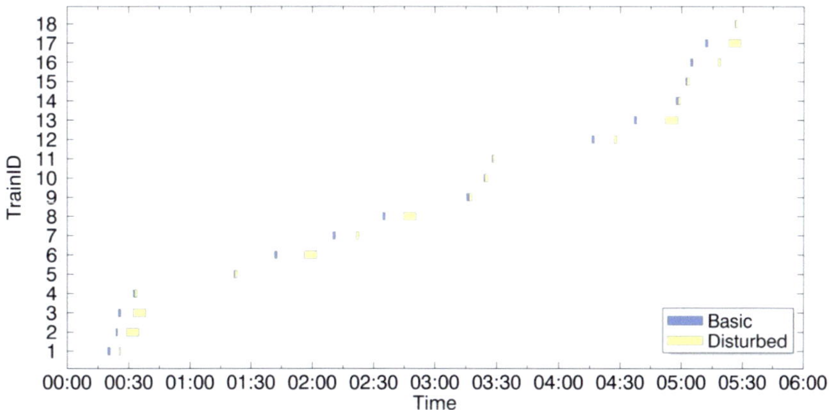

Figure 7-3   Arrival and departure time of all trains that operate on B8 in the investigation area with basic timetable (blue) and disturbed timetable (yellow), respectively

7.2.2   Simulation results of operational risk

Disturbance distribution types

Figure 7-4 shows the analysis of the simulation results by reporting *NRI* in the situation that a negative exponential distribution (Figure 7-4a) as well as an Erlang distribution (Figure 7-4b) is adopted for generating the imposed disturbances.

Results indicate that:

1) The basic timetable results in the smallest *TotalwWT* due to a lack of disturbance imposed on any block section;

2) Significantly larger *NRI* values are possessed by B6~B8, B11~B14, B19~B21, and B26~B28 than by other block sections, suggesting them as high operational risk block sections. In contrast, B18, B24, B29 and B32~B35 show minimal values of *NRI*, implying that the disturbances occurring on those block sections contribute almost no negative influence to the railway network;

3) A comparison between Figure 7-4a and Figure 7-4b display similar patterns of *NRI* values for all block sections, manifesting that different distributions of random disturbances have no significant influence on the simulation results. Figure 7-5 consists of a separate classification of each train class and depicts the average total weighted waiting time within each train class for all block sections. It is evident that although all block sections follow a similar trend with respect to the *NRI* shown in

Figure 7-4, there is significant variation regarding waiting time for different train classes. Freight transport (NGz) has a significant larger waiting time than long distance passenger transport (FRz) and regional passenger transport (NRz) due to the lower priority of NGz during railway operation compared with the other two train classes.

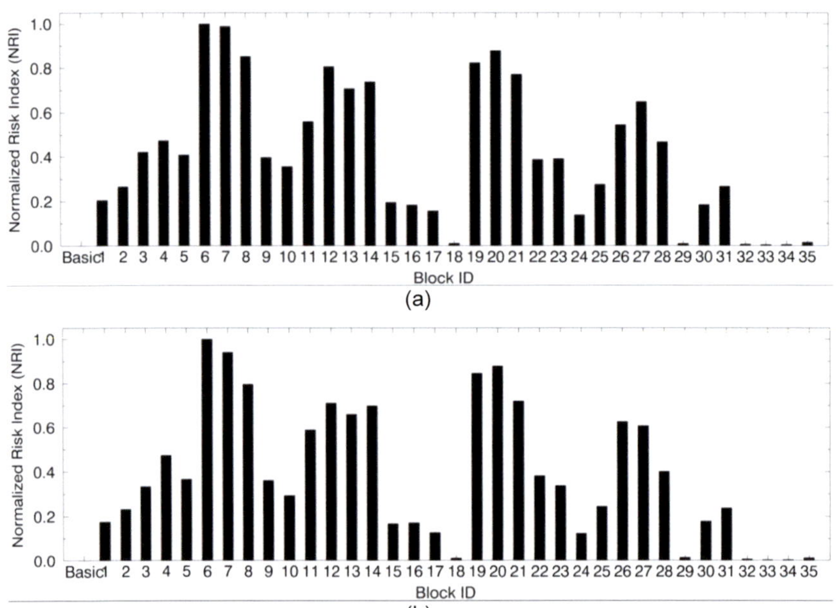

Figure 7-4   The normalized risk index (*NRI*, Normalized Risk Index) for each block section with simulated random disturbances following a) negative exponential distribution and b) Erlang distribution

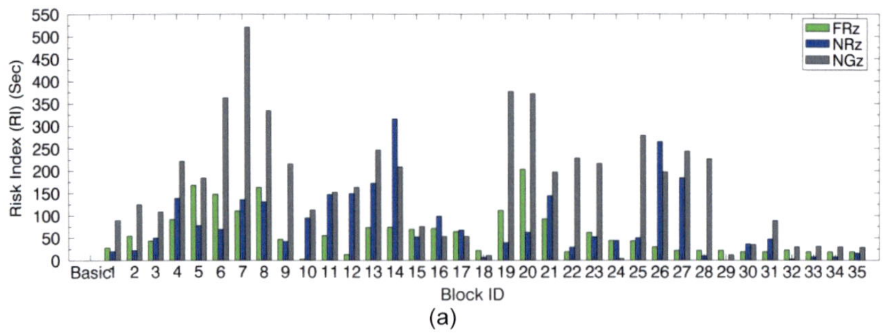

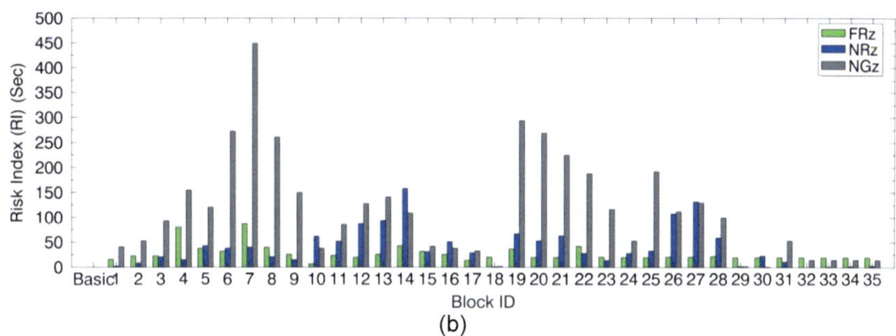

Figure 7-5   The RI of each train type for each block section with simulated random disturbances following a) negative exponential distribution and b) Erlang distribution

Traffic volume

The effects of the compression proportion of a basic timetable on the *RIs* and *NRIs* of various block sections are shown in Figure 7-6a. Block sections B6, B25, and B35 are selected as the representatives of high-, median-, and low-risk block sections, respectively. The *RIs* (average *TotalwWT* for basic timetable) of all block sections generally follow an increasing trend with the compression proportion. Although values of *RI* change along with compression proportion, the relative relationship among all three block sections is maintained at any proportion level. This characteristic is very important in explaining that relative *RIs* are basically independent to the total number of trains operating in the railway network, as long as the mixture of train types (structure of operating program) remains constant. Rather than the absolute value of *RIs*, Figure 7-6b displays the relationship between *NRIs* and the compression proportion of the basic timetable. It is evident from the figure that the *NRIs* of all three block sections show almost no change with an increasing compression proportion, further indicating that the traffic volume itself has very little contribution to the operational risk of block sections.

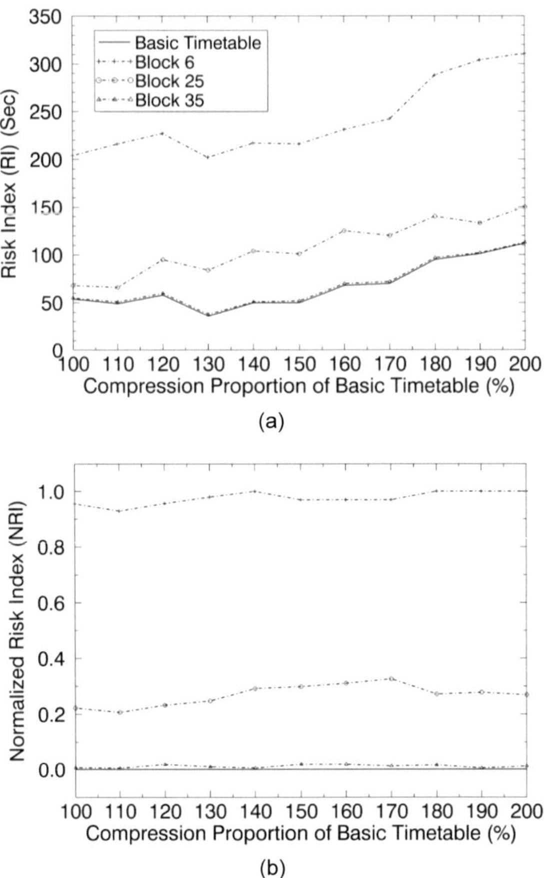

Figure 7-6   Variations of a) *RI* and b) *NRI* of the basic timetable and 3 representative block sections against the compression proportion of basic timetable

Freight transport schedules

In this section, the dwell time of NGz at train stations is adjusted from an initial 5 minutes (normal dwell time) to 15 minutes (extended dwell time), and meanwhile the departure time of certain freight trains are slightly postponed randomly. As manifested in Figure 7-7, increasing of NGz dwell time increases the *RI* (average total weighted waiting time) for each block section.

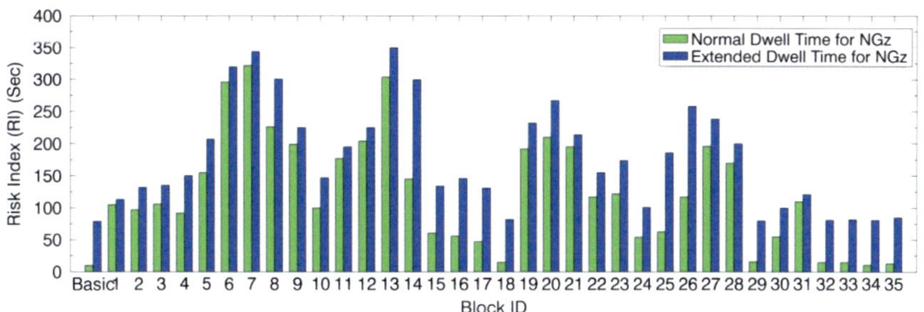

Figure 7-7   The RI of each block section in the case of normal and extended NGz dwell time in the basic timetable

Relative to one another, however, the values of *RI* among all block sections greatly remain constant, indicating that the values of *NRI* for each block section does not depend upon the slight adjustments made to the NGz in the basic timetable. Thus, the results of the operational risk level mapping for the studied railway network will not be influenced.

7.2.3   Determination of influencing parameters

Total number of operational risk level

An exceedingly low value of L_{total} (total number of operational risk level) results in too coarse of a classification to meet the requirement of some further investigations; while an overly large L_{total} will lead to significant variations of classification results among different basic timetables with different compression proportions. In order to achieve an appropriate balance, a trial-and-error test was conducted. By setting L_{total} from 1 to the 35 (the number of block sections in the investigation area) in turn, a relationship between the calculated *Averaged Nstd* and L_{total} can be obtained (see Section 3.4.1).

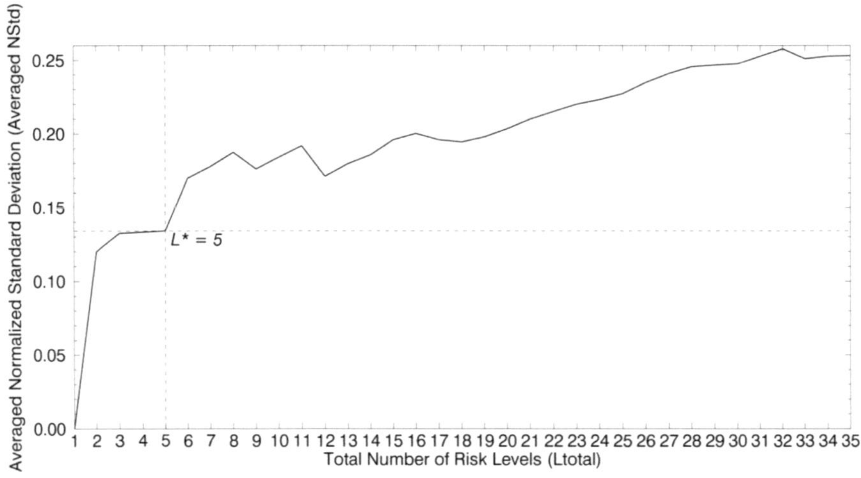

Figure 7-8   Relationship between the averaged normalized standard deviation (*Averaged Nstd*) of operational risk level and the selected total number of risk levels (L_{total})

The variation of *Averaged Nstd* with L_{total} is illustrated in Figure 7-8. Within the figure, a monotonic rise of the value of *Averaged Nstd* with L_{total} is observed, along with a drastic jump between L_{total} = 5 and L_{total} = 6. It can be inferred from this phenomenon that classifying all block sections into 6 (or higher) levels may not be reasonable when considering stability. In contrast, 5 levels of a classification scheme that is both stable and fine for the investigated reference example. In general, although the optimal levels usually varies with the specific structural characteristics of different railway networks, the same procedures stated in section 3.4.1 could be employed directly to find the best L_{total} accordingly.

Number of disturbance scenarios

In order to determine the appropriate number of disturbance scenarios, the test is conducted by increasing *NS* from 1 to 50 with an increment of 1. The relationship between *RIs* and *NS* is shown in Figure 7-9.

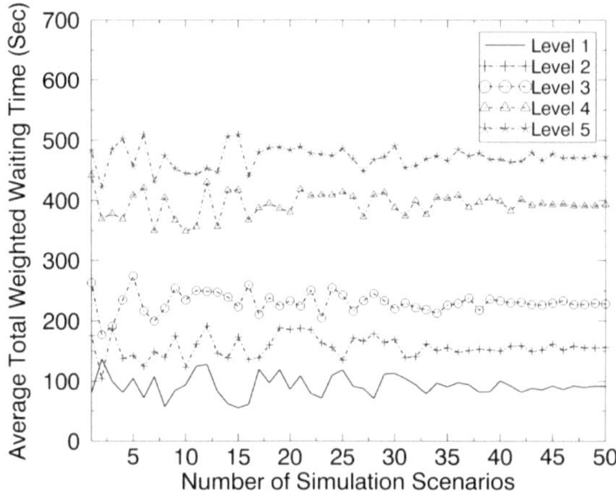

Figure 7-9   Relationship between the number of simulation scenarios and the average total weighted waiting time within each operational risk level

For small values of *NS*, it can be seen that the *RI*s of all the five risk levels vary greatly; while when *NS* reaches a value of about 40, the *RI*s become stable and the curves of the five risk levels shows obvious differences. As a result, an *NS* value of 40 for the investigated reference example is recommended within the current study. In theory, the convergence point of *NS* varies with railway networks due to different numbers of block sections and train runs. Considering that the operational risk analysis is an offline task which only needs to be performed once for a certain railway network, it is suggested, in practice, to conduct a trial-and-error test to determine the minimum *NS* that ensures the statistical reliability.

7.2.4   Operational risk classification and mapping

Operational risk mapping can be produced by firstly ranking all block sections based on the above calculated NRIs and classify them into different risk levels. In this case, the L_{total} =5. Accordingly, the classification results are given in Figure 7-10, and the final operational risk map is drawn in Figure 7-11.

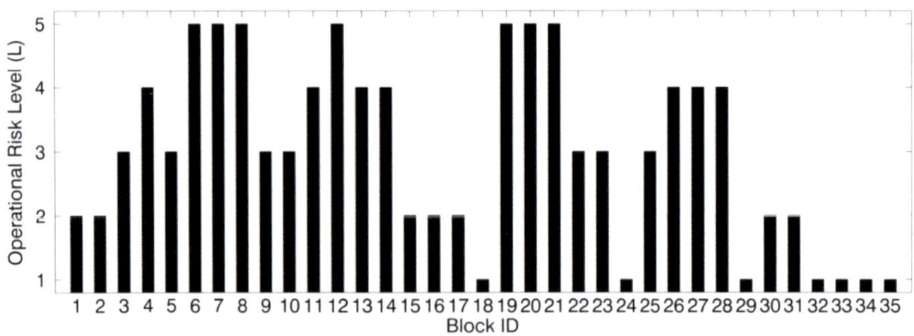

Figure 7-10   Statistics of operational risk levels of all block sections in the studied railway network (L_{total} =5)

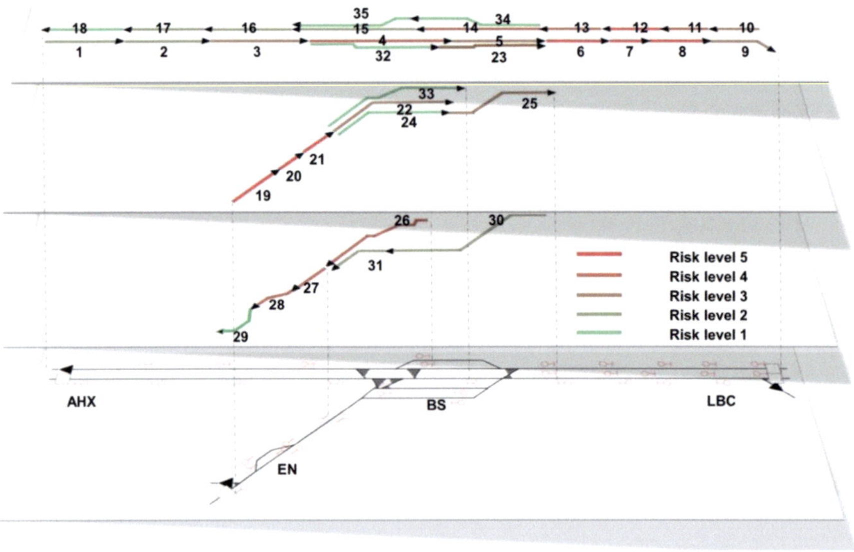

Figure 7-11   Operational risk map of all block sections in the studied railway network

In this example, the block sections with high operational risk (level 4 and level 5) are B4, B6~B8, B11~B14, B19~B21 and B26~B28, of which B6~B8 and B11~B14 have a high traffic volume, parts of B4, B19~B21and B26~B28 are the single-track block sections with trains operating in opposite directions. Specifically, B6~B8 are the block sections where trains from two directions (AHX-LBC and EN-LBC) merge. Compared to B26~B28, B19-B21 have higher operational risk level because not only the opposite direction (LBC-EN) but also other directions (AHX-LBC) will be influenced if operational disturbances occur. The block sections with low operational risk (level 1 and level 2) are B1, B2, B15~B18, B24 and B29~B35, which either have a low traffic volume or are the alternative train routes.

7.2.5   Summary of operational risk analysis

The proposed approach was verified with regards to its effectiveness on the reference railway network, within affordable efforts. Results indicate that the algorithm is capable of evaluating the operational risk level of block sections in a robust manner and maintaining compatibility with various distribution types of random disturbances.

In order to investigate the influence of the distribution types of random disturbances, the negative exponential and Erlang distributions were taken within this study as examples, and similar results were achieved. This result implies that the operational risks of block sections are mostly independent to the distribution type of disturbances. In fact, the proposed algorithm is capable of applying any sort of distribution type of disturbances. Researchers can define the distribution type based on the specific railway network or fit an empirical model if sufficient real data is at hand.

Operational risk analysis is the pre-study and important basis of the proposed dispatching algorithm. In practical, it is considered as an offline task which only needs to be conducted once for any specific railway network. Technically, the operational risk analysis is developed with C# in Visual Studio 2015 environment by calling the simulation tool RailSys® 7. It was compiled and executed on a Fujitsu computer (Intel Core i5-4670 CPU @ 3.40GHz, 8.00 GB of RAM). The operational risk analysis lasted 4.2 hours for a small reference railway network (in total 35 block sections, 40 scenarios for each block section). A decrease of computational time is possible through the implementation of the algorithm in a basic programming language (e.g., C), through the employment of parallel computing, or through the utilization of more efficient simulation tools.

7.3   Dispatching algorithm

In this section, the dispatching algorithm proposed in Chapter 4 is tested with the investigated railway network described in Section 7.1. The basic timetable created for this test contains the information of 36 trains operated on the investigated area (50% freight transport, 33.3% regional passenger transport, 16.7% long-distance passenger transport) within 6 hour time period. Corresponding to the four separate tasks of the dispatching algorithm shown in Figure 4-1, results of each task will be given in details in the following four sub-sections.

7.3.1　Stage division

In the rolling time framework (Figure 4-2), two key parameters for stage division are the prediction horizon (*PH*) and the dispatching interval (*DI*). Here, *PH* is set as 2 hours and *DI* is set as 0.5 hour for the test. It should be noted that the two parameters should be related to the stability of the studied railway network. For a stable network in which random disturbances rarely occur, a relative short *PH* and long *DI* are more appropriate to choose, and vice versa (the sensitivity analysis of these parameters will be conducted in Section 7.3). Given that *PH* = 2h and *DI* = 0.5h, Figure 7-12 shows the exact time points at which conflicts detection and dispatching are performed within the investigated time span.

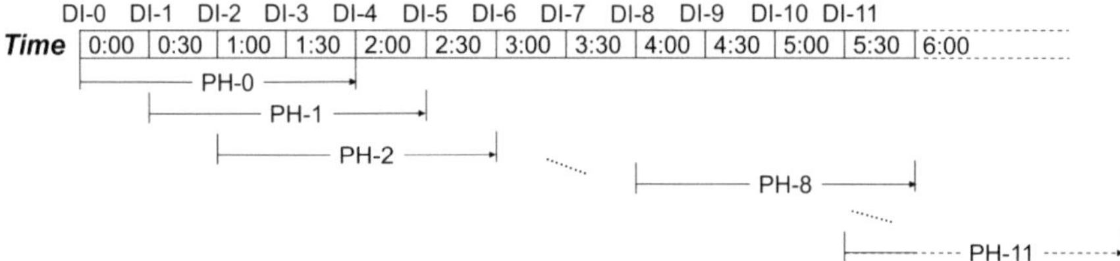

Figure 7-12　Stage division of the rolling time horizon framework with PH = 2h and DI = 0.5h. The investigated time span is 6 hours.

7.3.2　Risk-oriented conflicts detection at a single stage

Potential conflicts detection is performed time-driven at the beginning of each stage under the consideration of future random disturbances. As stated in Section 4.2, different amount of disturbances are artificially imposed on different block sections based on the operational risk level of those block sections, and the quantitative relationship between them was expressed in Equation 4-6. Considering that conflicts detection highly depends on the results of the operational risk analysis, the statistical distribution (negative exponential) and parameters of random disturbances used in this section are set as same as that in Table 7-1. Based on the operational risk analysis result given in Figure 7-10, all the block sections in the investigated railway network are classified into five risk levels. Consequently, the amount of four types of disturbances imposed on different block sections are calculated from Equation 4-6 (P_{max} and P_{min} are set as 0.6 and 0.05, respectively) and summarized in Table 7-2. It should be noted that although all values are shown in the table, entry delay is specific for the block section which is the entrance of the investigated area, and departure time extension and dwell time extension only applies for those block sections which contains a train station or a scheduled stop with a regular running time measurement point inside. At a certain *DI* point in Figure 7-12, the railway operation in the future *PH* period is simulated with RailSys® and the blocking time of each train on each block section are obtained. The modified blocking time is then calculated by adding the disturbances listed in Table 7-2 to the original blocking time. With that, track conflicts are identified if the modified blocking times of two or more trains overlap with each other in the same block section.

Table 7-2   Risk-oriented random disturbances imposed on different block sections in the process of conflicts detection. P_{max} and P_{min} in Equation 4-6 are set as 0.6 and 0.05, respectively. (unit: minute)

Operational risk level		L1	L2	L3	L4	L5
Entry delay	FRz	0.26	1.04	1.97	3.10	4.58
	NRz	0.10	0.42	0.79	1.24	1.83
	NGz	1.03	4.15	7.86	12.42	18.33
Departure time extension	FRz	0.05	0.21	0.39	0.62	0.92
	NRz	0.05	0.21	0.39	0.62	0.92
	NGz	0.26	1.04	1.97	3.10	4.58
Running time extension	FRz	0.05	0.21	0.39	0.62	0.92
	NRz	0.05	0.21	0.39	0.62	0.92
	NGz	0.51	2.08	3.93	6.21	9.16
Dwell time extension	FRz	0.05	0.21	0.39	0.62	0.92
	NRz	0.03	0.10	0.20	0.31	0.46
	NGz	0.26	1.04	1.97	3.10	4.58
Block sections IDs*		B18, B24, B29, B32, B33, B34, B35	B1, B2, B15, B16, B17, B30, B31	B3, B5, B9, B10, B22, B23, B25	B4, B11, B13, B14, B26, B27, B28	B6, B7, B8, B12, B19, B20, B21

***IDs of block sections in the investigated railway network are referred to Figure 7-11**

Three examples in Figure 7-13 shows the original and modified blocking time of trains operated on high (B8), medium (B3), and low (B18) operational risk block sections, respectively, during the *PH*-0 (0:00~2:00) period. According to the blocking time theory mentioned in Section 4.2:

1) Train T7, T1, T10, and T28 will potentially involve in severe track conflicts on block section B8 within *PH*-0 if random disturbances are taken into account in advance; while Train T25 and T13 will not likely be influenced by conflicts because of sufficient time interval in their original schedule;

2) Train T7 and T1 will probably encounter a conflict on B3, but the conflict is relatively slight;

3) No potential conflicts are observed on B18 even if random disturbances are considered.

In a similar way, the modified blocking time of all trains on all other block sections in the studied railway network can be obtained, and the objective of dispatching made at *DI*-0 is to solve all such detected conflicts and generate the rescheduled timetable for the *PH*-0 period.

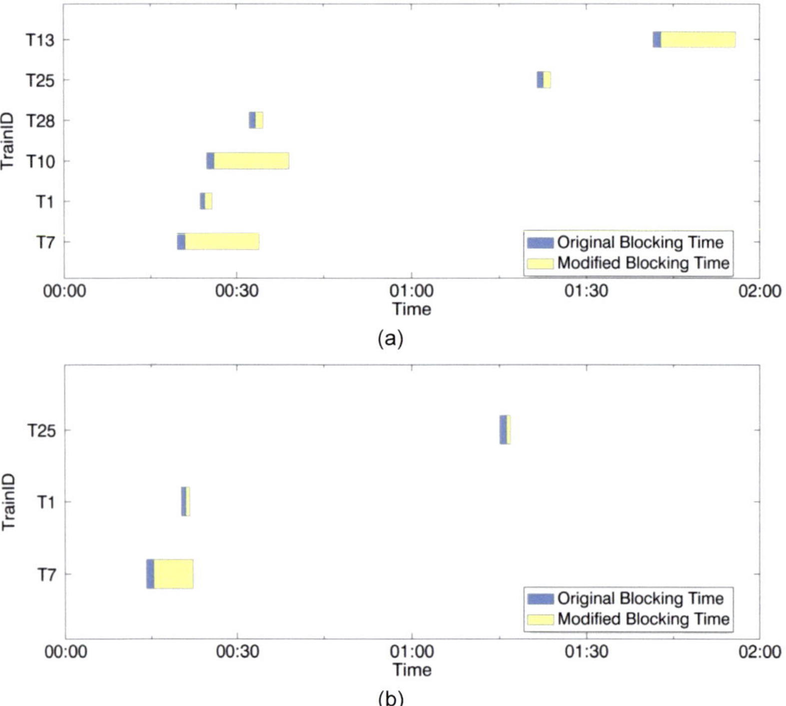

(a)

(b)

Hybrid Model for Proactive Dispatching of Railway Operation under the Consideration of Random Disturbances in Dynamic Circumstances

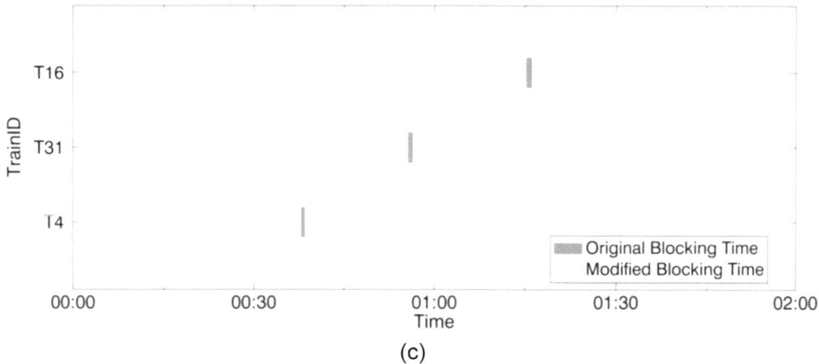

(c)

Figure 7-13   Original and modified blocking time of trains operated on block section a) B8, b) B3, and c) B18, respectively, during the *PH-0* period which starts from 0:00 and ends at 2:00.

7.3.3   Dispatching solutions at a single stage

With all conflicts detected for the entire railway network, the threshold *NT* of the *TotalwWT* described in Section 4.3.1 is used to distinguish slight conflict and significant conflict. In this case study, *N* is set as 5 minutes (300 seconds) for the test. Accordingly, the three examples listed in Figure 7-13 are identified as block sections with significant conflicts (B8), slight conflicts (B3), and no conflicts (B18), respectively, during *PH-0* period. For block section B8, the initial *TotalwWT* during *PH-0* is calculated as 6.5 min with modified blocking time implemented (see Figure 7-13 and Figure 7-14). Thus, reordering of conflicted trains (T7, T1, T10, and T28) using TS algorithm is at the first consideration.

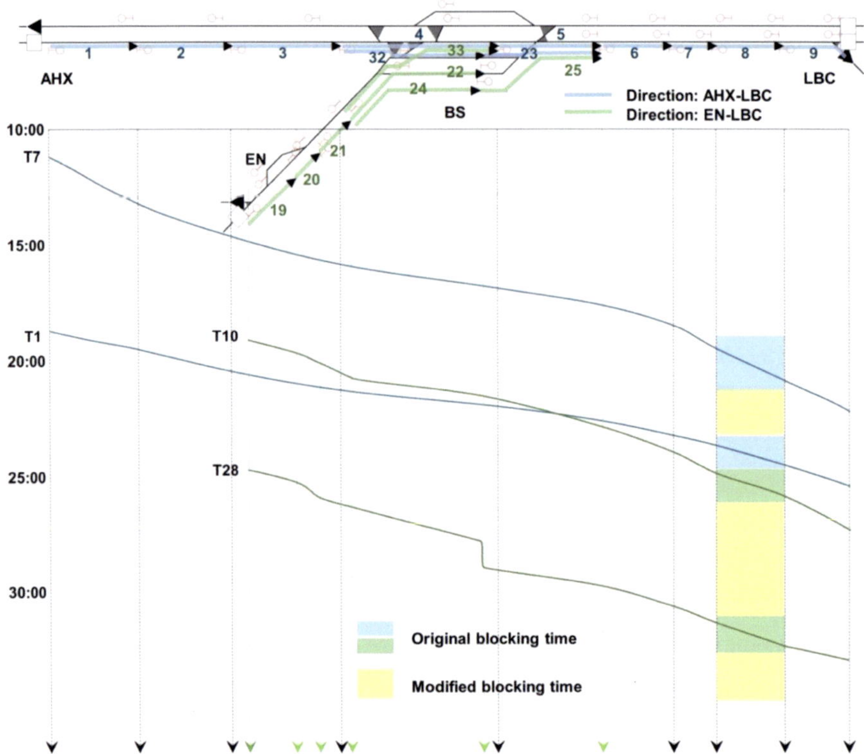

Figure 7-14   Train path conflict of T1, T7, T10, T28 on block section B8 during the *PH-0* period which starts from 0:00 and ends at 2:00.

Regarding the TS process shown in Figure 4-6, the termination condition is satisfied when search loops (N_{loop}) exceeds 30 (N_{loop} is set as 30 to achieve the balance between the performance of reordering solution and the computation time; refer to section 7.4.2.3 for details); and the aspiration criterion is met if the objective function $f(S)$ (*TotalwWT*) becomes smaller than 2 min in this test. Figure 7-15 displays the TS procedure by tracing the change of $f(S)$ against the train sequence passing the block section B8 during *PH-0* period. It is found that the TS procedure stopped at the end of 6th loop with $f(S) = 1.2$ min and the near-optimal dispatching solution as $S^* = $ [T1, T28, T7, T10]. In this example, while the termination condition has not been satisfied, the aspiration criterion was met as $f(S)$ reduced to be smaller than 2 min, indicating the final solution S^* is "good" enough and no further search was needed. For B3, the initial *TotalwWT* during *PH-0* is calculated as 3.1 min (< $N = 5$min), for which the slight conflict can be solved by simply rescheduling the departure time of T1 so that no overlaps exist between the blocking time of T1 and T7. For B18, no dispatching actions are needed because of no potential conflicts are observed between any two trains. In a similar way, reordering or retiming decisions can be made for all block sections, based on which the rescheduled timetable is generated for the entire railway network for the current stage *PH-0*.

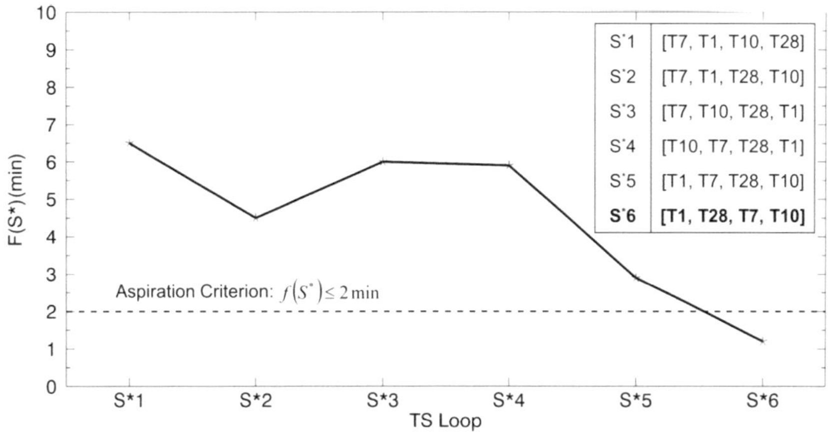

Figure 7-15   Change of the results of the objective function against the round of TS loop during *PH-0* period. Solution is specified as the train sequence passing the block section B8.

7.3.4   Dynamic dispatching in the rolling time horizon framework

Following the rules of rolling time framework stated in Section 4.4 and based on the stage division in this case study shown in Figure 7-12, the railway operation will stick to the rescheduled timetable generated at 0:00 till the beginning of *PH-1* (at 0:30). Then, conflicts detection will be performed again to check if potential conflicts exist during *PH-1* (from 0:30 to 2:30). Dispatching solution for *PH-1* will be also generated if conflicts detected, and will be stuck to till the beginning of *PH-2* (at 1:00). This iterative process will not be terminated until the end of the investigation time span (at 6:00). RailSys® is taken as the surrogate of real-world practice to simulate the railway operation. Meanwhile, to reflect reality as more as necessary, random disturbance samples generated in a Monte-Carlo scheme based on negative exponential distribution with parameters listed in Table 7-1 are imposed on all block sections. As results, Figure 7-16 traces if and what dispatching actions were implemented for each stage during 0:00~6:00. For *PH-0*, *PH-4*, and *PH-8*, significant potential conflicts were detected and thus reordering dispatching actions were taken for these three stages; for *PH-3* and *PH*-9, slight potential conflicts were detected and retiming was enough to solve; for all other stages, no conflicts were observed and no dispatching actions were needed.

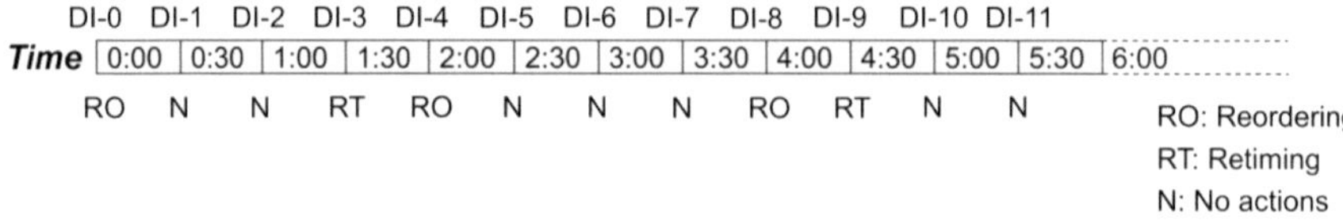

<pre>
 DI-0 DI-1 DI-2 DI-3 DI-4 DI-5 DI-6 DI-7 DI-8 DI-9 DI-10 DI-11
Time | 0:00 | 0:30 | 1:00 | 1:30 | 2:00 | 2:30 | 3:00 | 3:30 | 4:00 | 4:30 | 5:00 | 5:30 | 6:00
  RO   N   N   RT   RO   N   N   N   RO   RT   N   N   RO: Reordering
                                                                                RT: Retiming
                                                                                N: No actions
</pre>

Figure 7-16   Dispatching solutions generated for each stage during the investigation time span covering 6 hours.

7.3.5   Summary of dispatching algorithm

The proposed dispatching algorithm is implemented on the same railway network as that in the operational risk analysis. The prediction horizon (*PH*) and dispatching interval (*DI*) are set as 2 hours and 0.5 hour, respectively, for the test. Potential conflict detection is performed at the beginning of each stage based on the operational risk level of block sections. With that, cases of significant conflicts, slight conflicts, and no conflicts (e.g., Figure 7-13) are identified with respect to specific *PH* periods on specific block sections. Accordingly, reordering calculated from TS approach is carried out for significant conflicts while retiming is taken for slight ones. Within the rolling time horizon framework, conflict detection and dispatching are performed dynamically and iteratively till the end of the investigated time span.

In contrast to the operational risk analysis settled as an offline task, the conflict detection and dispatching are considered as online tasks in practical which should be conducted in actual railway operation. On the technical side, the dispatching algorithm is also developed with C# on the same platform (Fujitsu computer, Intel Core i5-4670 CPU @ 3.40GHz, 8.00 GB of RAM). For any stage, it takes about 20 seconds in total to complete the calculation of both conflict detection and resolution, which could be further improved if basic programming language (e.g., C) or higher-performance simulation tools are utilized. Although the offline process take a relatively long while (4.2 hours; refer to section 7.2.5), the high efficiency of the online process is quite enough to ensures the practical use of the proposed dispatching algorithm.

7.4   Assessment of the dispatching algorithm and sensitivity analysis of dispatching-related parameters

In section 7.3, sample results of specific block section and specific trains were provided to illustrate the proposed dispatching algorithm. In this section, statistics for the entire railway network will be presented and analyzed to evaluate the algorithm. First, the three indicators described in Section 5.1 are calculated and compared; then, results of sensitivity analysis of those dispatching-related parameters are reported, respectively.

7.4.1   Assessment of dispatching algorithm over stages

Figure 7-17 records the calculated *TotalwWT* over dispatching stages. It is found that the *TotalwWT* decreases over stages, indicating that, with the dispatching algorithm implemented, most of potential conflicts are solved in earlier stages, resulting in higher punctuality in the later stages even if random disturbances still occur.

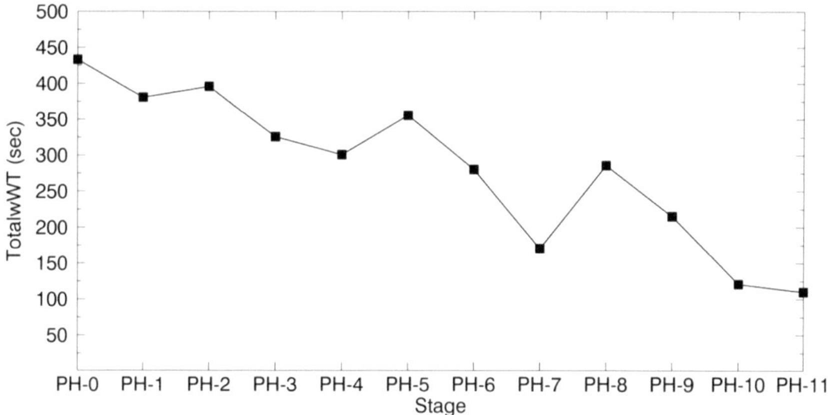

Figure 7-17   The *TotalwWT* of each stage calculated from railway simulation results

Figure 7-18 illustrates how *NRR* (number of relative reordering) and *AAR* (average absolute retiming) vary with stages under the circumstance of dispatching algorithm carried out during the railway operation.

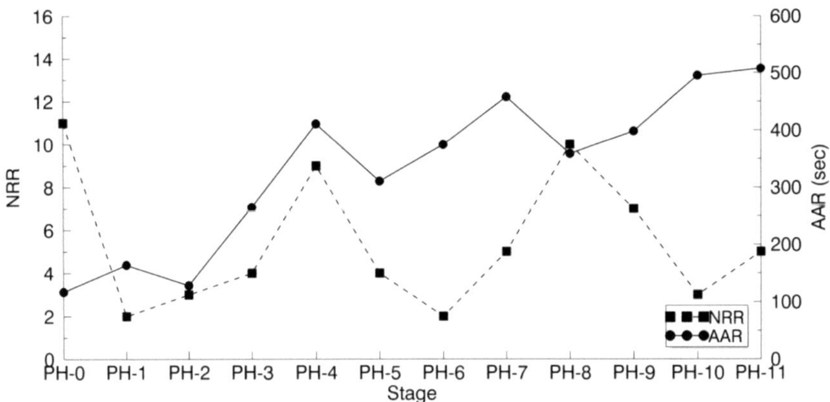

Figure 7-18   The *NRR* and *AAR* for each stage calculated from railway simulation results

The general trend of *NRR* is that a relatively large *NRR* is followed by several relatively small *NRR*. In the proposed algorithm, the dispatching solution is calculated for a future *PH* period, while the dispatching interval *DI* is smaller than *PH*. A large *NRR* suggests "significant" dispatching which will maintain the railway network as states of slight conflicts in future *PH* period, and thus several immediate following stages will not be detected as significant conflicts. For *AAR*, it has an obvious increasing trend over stages, which is reasonable considering that the retiming during dispatching postponed train schedules more significantly at later stages than that of earlier stages.

7.4.2   Sensitivity analysis of dispatching parameters

In Section 7.3 and 7.4.1, dispatching results were provided and evaluated with specific dispatching parameters (*PH* = 2h, *DI* = 0.5h, P_{max} = 0.6, and N_{loop} = 30). In this section, results of the sensitivity analysis of each parameter will be reported based on the method described in Chapter 6.

PH (prediction horizon) and DI (dispatching interval)

In the case of different combinations of *PH* and *DI*, railway operation is simulated by implementing the dispatching algorithm. Table 7-3 summarizes the overall *TotalwWT* calculated from the simulation results and Figure 7-19 displays those statistics more clearly. It is found that:

1) For a fixed *PH*, *TotalwWT* increases with *DI*, suggesting that performing conflicts detection and dispatching at higher frequency can handle random disturbances in time and thus leads to lower *TotalwWT*;

2) For a fixed *DI*, *TotalwWT* decreases with *PH*, suggesting that the dispatching solution with longer future period taken into account is more robust and more effective on solving the conflicts

Based on the analysis, in theory, combination of the maximum *PH* (100% investigation time span T_E) and the minimum *DI* (5% *PH*) generates the best railway operation. However, involving the entire investigation time span into consideration for each round of conflicts detection and dispatching calculation would induce pretty high computation burden.

From Table 7-3 and Figure 7-19, it can be seen that within the whole *DI* range, the changing (decreasing) trend of *TotalwWT* becomes gentler at higher *PH* value region, indicating that *PH* plays more important role on the dispatching quality when itself is small, and vice versa. Accordingly, a relatively short *PH* is also able to ensure the dispatching quality without significantly enlarging the *TotalwWT*. Specifically in this case study, *PH* within 30%~40% of T_E and *DI* within 20%~25% of *PH* are recommended.

Table 7-3　　The overall *TotalwWT* (unit: second) of the railway operation with dispatching algorithm implemented for different combinations of *PH* and *DI*. T_E is 6h in this case study.

DI (% of PH)		PH (% of T_E)									
		10	20	30	40	50	60	70	80	90	100
	5	467	393	344	297	257	309	318	306	194	252
	10	510	439	348	326	316	332	265	265	238	229
	20	622	505	362	300	266	332	320	257	311	312
	25	665	605	438	323	294	259	242	258	273	322
	50	794	641	532	484	446	362	381	324	278	278

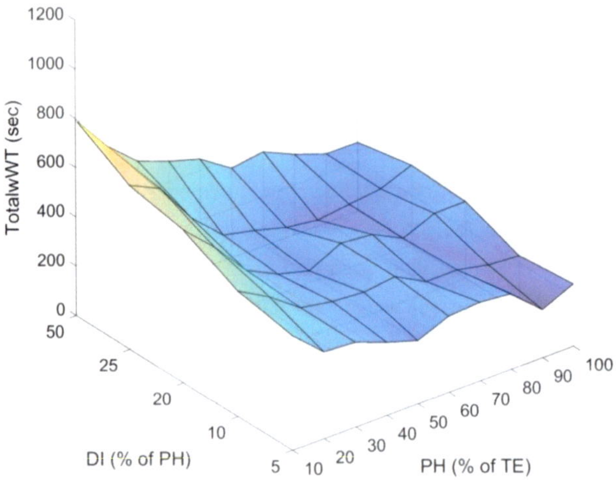

Figure 7-19   Variation of the *TotalwWT* with *PH* and *DI* based on the data given in Table 7-3.

Similar to *TotalwWT*, variations of *NRR* and *AAR* against *PH* and *DI* are plotted in Figure 7-20. Note, that values shown here are the averaged values over all stages. It is observed in this investigation area that:

1) For a fixed *PH*, averaged *NRR* decreases with *DI* very slightly, suggesting that the frequency of dispatching has almost no influence on the reordering times in dispatching solutions once *PH* is specified;

2) For a fixed *DI*, averaged *NRR* obviously increases with *PH*. This is reasonable because a relatively larger *PH* spans a longer future period and involves more trains into consideration, which probably detect more potential track conflicts and thus more reordering is needed, and vice versa;

3) Within the whole *DI* range, the increasing rate of averaged *NRR* increases with the value of *PH*, indicating that the influence of *PH* on *NRR* becomes larger when itself is large;

4) Averaged *AAR* fluctuates slightly with the changing of *PH* or *DI*, implying that the average deviation over stages between basic timetable and rescheduled timetable has almost no relationships with *PH* or *DI*.

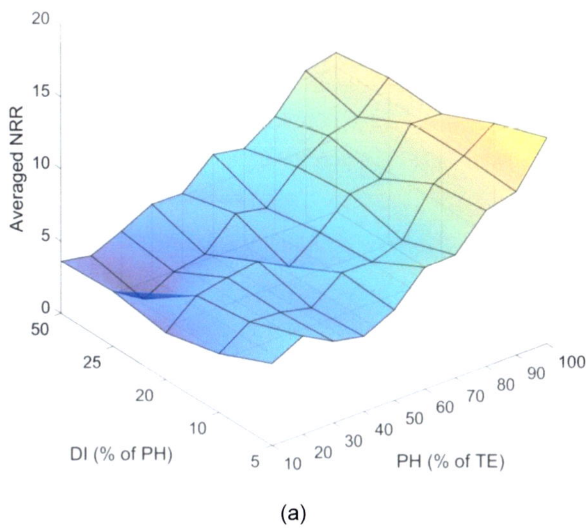

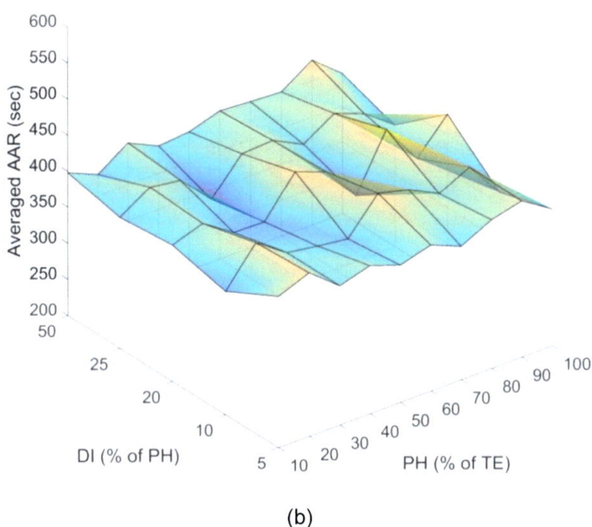

Figure 7-20   Variation of the a) *NRR* and b) *AAR* against *PH* and *DI*.

Based on these analysis, on one hand, a short *PH* corresponds to a lower *NRR* that brings less computation burden in the price of degraded punctuality (larger *TotalwWT* as shown in Figure 7-19); on the other hand, a long *PH* may induce widely reordering in the railway network, although the operation quality is quite ensured (lower *TotalwWT* as shown in Figure 7-19). Figure 7-20a gives clue to achieve such a balance. It is seen that *PH* around 40% of T_E locates at the critical point where the *NRR* starts to increase drastically with *PH*, which is consistent with recommended *PH* value accordingly to Figure 7-19.

Pmax (maximum value of cumulative probability regarding disturbances distribution)

P_{max} was set to be varying from 0.30 to 0.95 with interval of 0.05 to test the dispatching algorithm. During the test, *PH* and *DI* were set as the recommended values (*PH* = 40%T_E ≈ 2.5h; *DI* = 20%*PH* = 0.5h) described in Section 0. According to the conflicts detection approach described in Section 4.2, a small P_{max} results in inadequate consideration of potential random disturbances, which probably degrades the effects of dispatching solutions; while a too large P_{max} means excessive inclusion of disturbances which lead to redundant time space in the rescheduled timetable that may deteriorate the railway capacity. To find the most appropriate P_{max} to achieve the balance, Figure 7-21 displays how the *TotalwWT* change with P_{max}.

1) As expected, the statistics demonstrate that the relationship between *TotalwWT* and P_{max} changes from decreasing trend to increasing trend with P_{max}, and *TotalwWT* achieves the lowest value around the P_{max} range of [0.55, 0.7];

2) When P_{max} is smaller than 0.55, the more consideration of potential disturbances contributes to a higher punctuality of the railway operation with dispatching implemented;

3) If P_{max} reaches up to 0.7 or larger, *TotalwWT* increases with P_{max} dramatically, implying that reordering and retiming are not able to handle the imposed disturbances. Train schedules have to be postponed significantly.

Based on the analysis, in this case study, the most appropriate P_{max} falls within the range of [0.55, 0.7]. Further, considering that a larger P_{max} generally increases the computation burden, P_{max} is recommended as 0.55 to achieve the balance between the dispatching effects and the computation time.

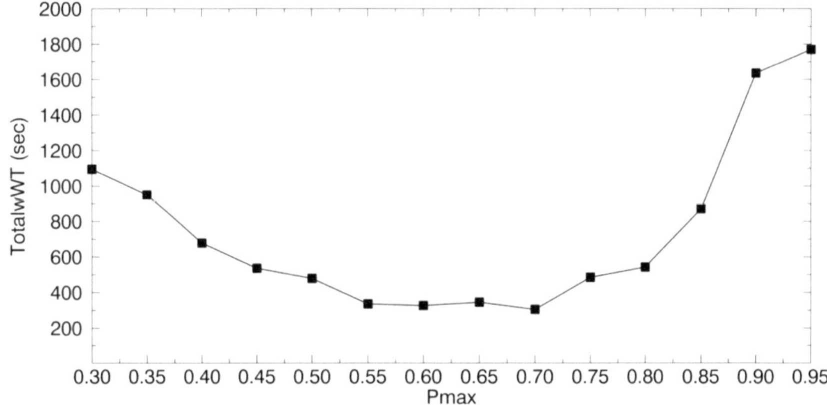

Figure 7-21   The *TotalwWT* calculated from the simulation results in the case of differenct P_{max}.

Nloop (Search loop limit of Tabu search algorithm)

Tabu search algorithm is a local optimization method in which the termination condition N_{loop} has direct influence on the final solution found. A too small N_{loop} usually doesn't ensure the quality of the solution, while a too large N_{loop} is waste of computation time without contributing to obtain a much better solution. To find the most appropriate N_{loop}, different values were set to test the dispatching algorithm and results are given in Figure 7-22. Generally speaking, *TotalwWT* decreases and converges with N_{loop}. For a relatively low P_{max} (0.3), the convergence point of N_{loop} (around 10) is smaller than those (around 30 or 70) in the case of a medium (0.55) or a high (0.95) P_{max}, respectively, which is attributed to that the larger amount of disturbances imposed, the more difficult to find the optimal reordering solution to solve the conflicts. Furthermore, Figure 7-23 quantifies the convergence point of N_{loop} in the case of each P_{max}. Accordingly, based on the optimal P_{max} decided as 0.55 in section 7.4.2.2, N_{loop} is thereby recommended to be set around 30 to 35 to achieve the best balance in the case study.

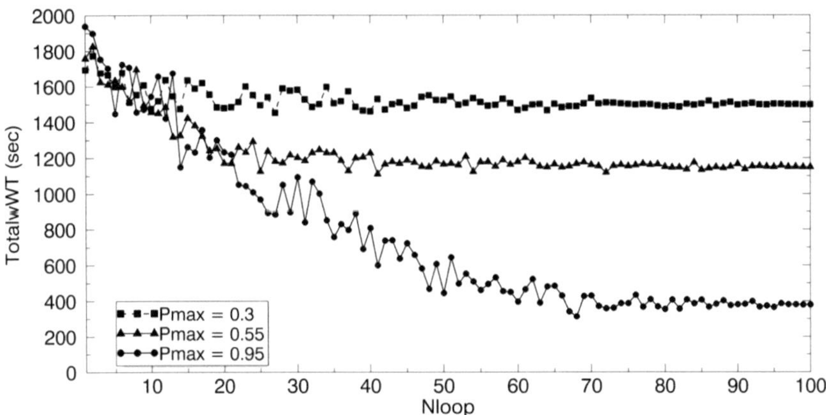

Figure 7-22   Change of the *TotalwWT* against N_{loop} in the case of a low and a high P_{max}, respectively.

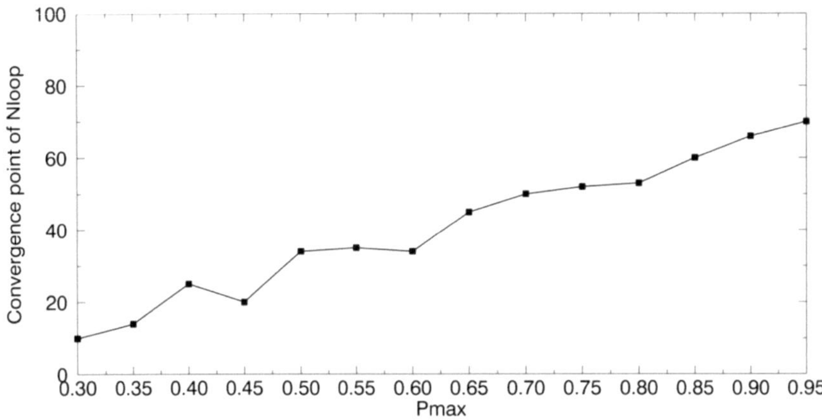

Figure 7-23   Relationship of the convergence point of N_{loop} with P_{max}.

7.4.3   Comparison with FCFS dispatching principle

In order to manifest the advantage of the proposed algorithm in this study, the widely used FCFS dispatching principle is also utilized on the same railway network. During this test, the recommended parameters as analyzed in section 7.4.2 are employed (i.e., $PH = 40\% T_E \approx$ 2.5h, $DI = 20\% PH = 0.5$h, $P_{max} = 0.65$, and $N_{loop} = 45$). Figure 7-24 reports the *TotalwWT* of each stage in the case of rescheduled timetable with the proposed algorithm implemented, and rescheduled timetable with FCFS implemented, respectively. Similar to Figure 7-17, both curves decrease over time passing due to the resolution of potential conflicts in earlier stages.

However, the effects on the reduction of *TotalwWT* from FCFS are obviously weaker than that of the proposed algorithm. These statistics further demonstrating the effectiveness of the proposed algorithm for ensuring the overall punctuality of the railway network.

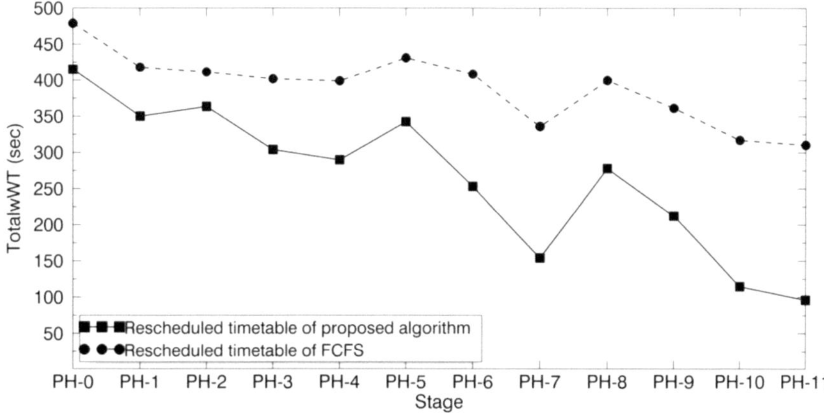

Figure 7-24   The *TotalwWT* of each stage calculated from railway simulation results of the reschdueld timetable with the proposed dispatching algorithm implemented and the FCFS dipatching principle implemented, respectively.

8 Specification of the developed algorithm

In order to avoid the error-prone informal text formalization of the proposed dynamic dispatching algorithm, a *Unified Modeling Language (UML)* is introduced in this chapter, which captures the architecture of the actual implementation. The employed module of core model (see Section 8.1), the requirements (see Section 8.2), the application scenarios (see Section 8.3) and the system architecture (see Section 8.4) are specified in detail for the application of the proposed dynamic dispatching system automatically.

8.1 Core model structure

During the process of dispatching algorithm assessment and dispatching-related parameters sensitivity analysis, an integrated core model developed by *Institute of Railway and Transportation Engineering (IEV)* is employed. The whole structure of the core model is shown in Figure 8-1.

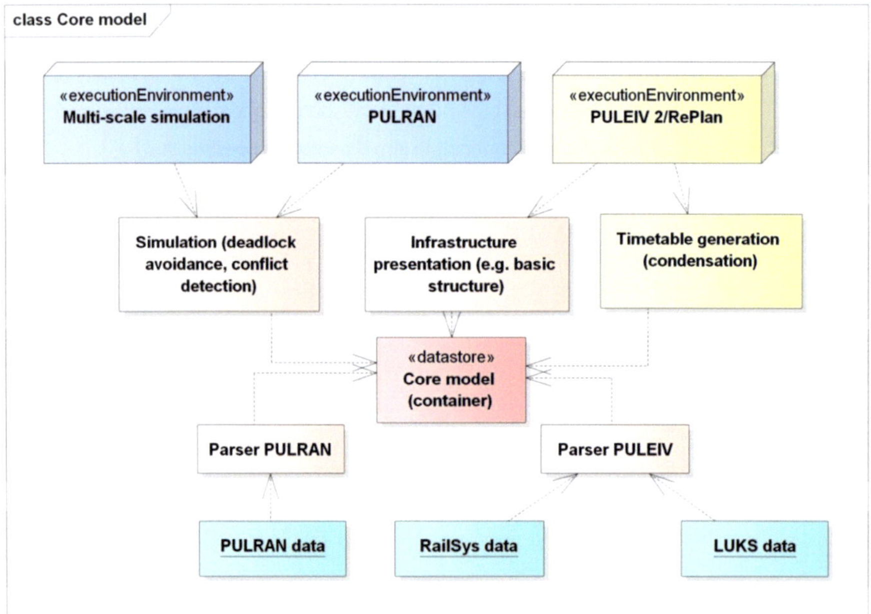

Figure 8-1   Structure of the core model developed by IEV

PULRAN data, RailSys data and LUKS data, they are all written in xml/csv format files. Among them, RailSys data and LUKS data are similar, which consist of three parts: infrastructure data, rolling stock data and schedule data. It can be imported directly from DB data format (xml-KSS and xml-ISS). As for PULRAN data, besides the data of infrastructure network, train group definitions and operational concepts, the occupancy information of each infrastructure resources are included. This core model provides a standard data store pattern in the form of internal object model, which can easily transfer the data type from other software including PULRAN, RailSys and LUKS automatically. For this purpose, several parsers are provided for data formats, including RailSys, railML, ISS (infrastructure data), KSS (timetable data), as well as other internal data formats applied in IEV. In this manner, the core model-based algorithms are not restricted to any employed software. Tasks can be fulfilled independent from applied simulation tools.

Based on the core model, three main functions including simulation (deadlock avoidance, conflict detection and resolution), infrastructure division and presentation, timetable condensation can be realized through different execution environments. Among them, multi-scale simulation is developed for conducting dispatching with micro-, meso-, macro-scale or the combination of them to deliver solutions quickly with sufficient performance in (Martin 2014b). *PULRAN (Program to Research the Logistic in the Shunting Operation)* aims at implementing and evaluating the railway synchronous simulation model as well as the Banker's algorithm (deadlock avoidance and prevention). PULEIV 2/RePlan provides a platform to evaluate the operation quality using indicators delay coefficient and punctuality degree, locate the bottleneck (bottleneck relevance, bottleneck significance and potential reserves) and investigate the capacity of the railway network (recommended area of the traffic flow).

In this study, the module of timetable condensation on the available platform of PULEIV 2 (marked by yellow) is employed, so as to free the applicability of assessment and sensitivity analysis function of the proposed algorithm

8.2   Functional and non-functional requirements

The requirements can be divided into functional and non-functional requirements. Typically, functional requirements include information about all the implemented system modules as well as their surroundings. They provide the broad audience with the intended behavior of the system. In this study, they are stated as the implemented functionality of the module, e.g., the use case (see Section 8.3) is built depending on the functional requirements of the module. In this case, the functional requirements of the proposed dispatching system are:

1) Operational risk analysis of the involved block sections

   The operational risk index and the normalized operational risk index are calculated. The block sections in the investigation area are differentiated with different operational risk levels and the operational risk map is depicted according to their vulnerability to the stochastic disturbances (see Chapter 3).

2) Provision of the dynamic dispatching solutions in the circumstances of conflicts

   The optimal or near-optimal dispatching solutions are provided automatically for every prediction horizon and updated after every dispatching interval based on the hybrid dispatching model with combination of heuristic and simulation approaches (see Chapter 4).

3) Assessment of the proposed dynamic dispatching algorithm

   The proposed algorithm and the generated dispatching solutions are evaluated and validated based on several indicators (e.g., TotalwWT, NRR, AAR). Comparison between the proposed algorithm and FCFS is conducted to testify the effectiveness of the developed algorithm (see Chapter 5).

4) Sensitivity analysis of dispatching-related parameters

   The sensitivity analysis of dispatching-related parameters, such as PH, DI and Pmax, is performed. Parameters are adjusted in the process in order to investigate the influence of these dispatching-related parameters on the performance of railway operation for the considered studied area (see Chapter 6).

The non-functional requirements, which characterize the quality attributes of the dynamic dispatching system, must be considered in the design of the system architecture (see Section 8.4). The non-functional requirements are:

1) Independence from the stochastic disturbances *PDF*

   The dynamic dispatching system should be valid for different probability distribution functions of the stochastic disturbances and shows no performance variance among different stochastic PDF.

2) Independence from the infrastructure and the operating program

   The usability of the proposed dynamic dispatching system should be restricted to neither the concrete infrastructure nor the fixed operating program. The procedures of implementing the algorithm will be the same with an extended infrastructure or a congested operating program.

3) Independence from the concrete simulation tool

   The employed simulation software for representing the operation of the real world can be substituted. The sequence of the applied algorithm has no knowledge regarding to a specific simulation tool. The components which implemented in the dispatching system will not be influenced if the simulation software is changed.

4) Extensibility and changeability

   Because of the various dispatching objectives regarding to different interest groups, the considered evaluation indicators can be different. For this purpose, the dynamic dispatching system provides the users flexible indicators configuration.

5) Low calculation effort

   The calculation efforts should be significantly lower than the efforts with manual dispatching. Here not only the time expenditure but also the conducting complexity should be considered. For instance, an automated generation package of dispatching solutions has the advantage over the same time expenditure with much lower labor consuming.

There are also other non-functional requirements, such as reliability, applicability, and security. Most of them are the IT implementation details and technology with no direct relevance of this thesis. Therefore, they are not discussed in this thesis.

8.3   Application scenarios

According to the functional requirements, the three application scenarios of the proposed dynamic dispatching system are specified in Figure 8-2.

All the application scenarios mentioned below employ an external application "Implement the simulation", which is not within the system boundary of the dynamic dispatching system. Thus, the details of a specific simulation tool are isolated from the dynamic dispatching system.

A typical workflow of the dynamic dispatching system begins with the "operational risk analysis" application. Thereafter, the dispatching solution generator is conducted based on the innovative conflicts detection results with consideration of future risk-oriented random disturbances. The dispatching-related parameters are calibrated afterwards to determine the appropriate parameters under the condition of the applied infrastructure and operating program. Finally, the validation process is performed between the proposed and the conventional dispatching algorithm.

The activity diagrams depicts the control flowing from one activity to another including the logic of conditional structures, loops as well as concurrency of each application scenario.

Figure 3-1 shows the procedure for "operational risk analysis" application. Figure 4-8 demonstrates the workflow for "dynamic dispatching" application. Workflows for "single stage" and "Tabu Search" are further presented explicitly in Figure 4-5 and Figure 4-6. The activity diagrams are not repeated in this chapter, although these diagrams are the most important artifacts of the specification.

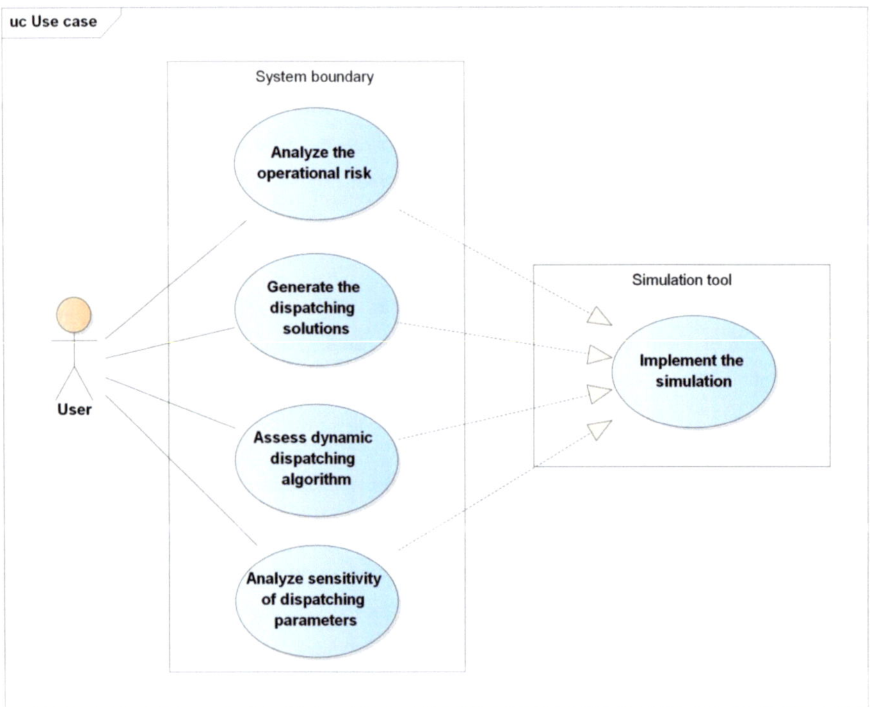

Figure 8-2   Use case diagram

8.4   Architecture of the dynamic dispatching system

The architecture of the dynamic dispatching system is developed based on the multiple and concurrent "4+1 views" with concern of various stakeholders from different perspectives stated in (Kruchten 1995). The "+1" view is a complementary view (see Section 8.3, application scenarios) in order to illustrate and validate the system after the architecture design is finished. The other 4 different views including logical view, process view, physical view and development view are given below:

1) <u>The logical view</u>, which represents the object model, is depicted by a conceptual model with components structure in Section 8.4.1.

2) <u>The process view</u>, which focuses on the interaction between objects, is described by a sequence diagram with functions and return values in Section 8.4.2.

3) <u>Development view</u>, which concentrates on the static organization of the system in its development environment, is demonstrated by a component diagram in Section 8.4.3.

4) <u>Physical view</u>, which reveals the mapping relationship between software and hardware. In this study, only one thread of one computer is employed for the implementation of the dynamic dispatching system. Therefore, a deployment diagram which illustrates the architecture of hardware devices, servers (e.g., web server, application server, and database server) and software execution environment is not necessary. The calculation performance of this dispatching system regarding to the applied computer will be discussed in Section 9.1.

8.4.1   Logical view

Figure 8-3 illustrates the logical view of the proposed dynamic dispatching algorithm.

The class "Disturbed Timetable Generator" is responsible for the generation and configuration of the disturbances scenarios. Generally speaking, each disturbance scenario corresponds to one specific disturbed timetable. There are 3 different types of disturbed timetables considered for different objectives in this system, which are all generated based on the class "Basic Timetable" and "Disturbance Distribution".

1) Function "DisturbedTimetableBlock"

This function aims to generate disturbance scenarios regarding to only one specific block section (target block section), applied in operational risk analysis module. In Chapter 3, in order to evaluate the operational risk of a certain block section, for each train that passes through the investigated block section, random disturbances are imposed on the train when it occupies that block section (no disturbances are imposed on trains when they occupy other block sections).

Multiple disturbance scenarios will be generated and simulated for the same target block section, so as to obtain the average value of all scenarios regarding to the related indicator (TotalwWT), deemed as the operational risk of the target block section. By repeating this procedure for each block section one by one, the operational risk map of the whole railway network can be obtained.

2) Function "DisturbedTimetable"

The aim of this function is to generate the disturbance scenarios more likely to be close to reality, so as to provide a proxy test field for continuously updating the railway operation and for evaluating the performance of the proposed algorithm like in reality.

Any train on any block section probably encounters random disturbances in reality. Therefore, the disturbed timetable generated by this function contains disturbances on every train when they occupy every block section (not just the target block section). In addition to deriving the disturbances samples from known statistical distribution models, it can also be collected from empirical data in reality.

Another difference with the function "DisturbedTimetableBlock" is that no multiple disturbance scenarios are generated for the railway simulation of the same stage, because in reality, there is a unique value of a disturbance for a certain train on a certain block section at a time, which corresponds to only one random sample generated from this distribution function. In other words, the function "DisturbedTimetableBlock" is similar to an off-line task which is allowed to make use of sufficient historical data to investigate the operational risk of each block section in the railway network; while the function "DisturbedTimetable" is similar to an on-line task which is a simulation of the real world, and thus only one possibility of an disturbed timetable exists at a time.

3) Function "DisturbedTimetableArti"

The third function "DisturbedTimetableArti" is to artificially impose disturbances in the basic timetable on all block sections based on their different operational risk level, in order to provoke the "Conflict Detection" class. Different from the first 2 disturbances scenarios, the values of disturbances samples are not directly derived from known distribution models but on the basis of the model calculated by a mathematical formula with one parameter L (Operational risk level) (see Section 4.2). With consideration of the parameter L, "DisturbedTimetableArti" is able to only concentrate on the forthcoming significant disturbances for "Conflict Detection" class.

The tasks of classes "reordering" and "retiming" are for the execution and the configuration of the dispatching process. During the process, the interface "RailSys Simulation" and "Dispatching" interacts with each other iteratively for handling the dynamic and stochastic condition in the system.

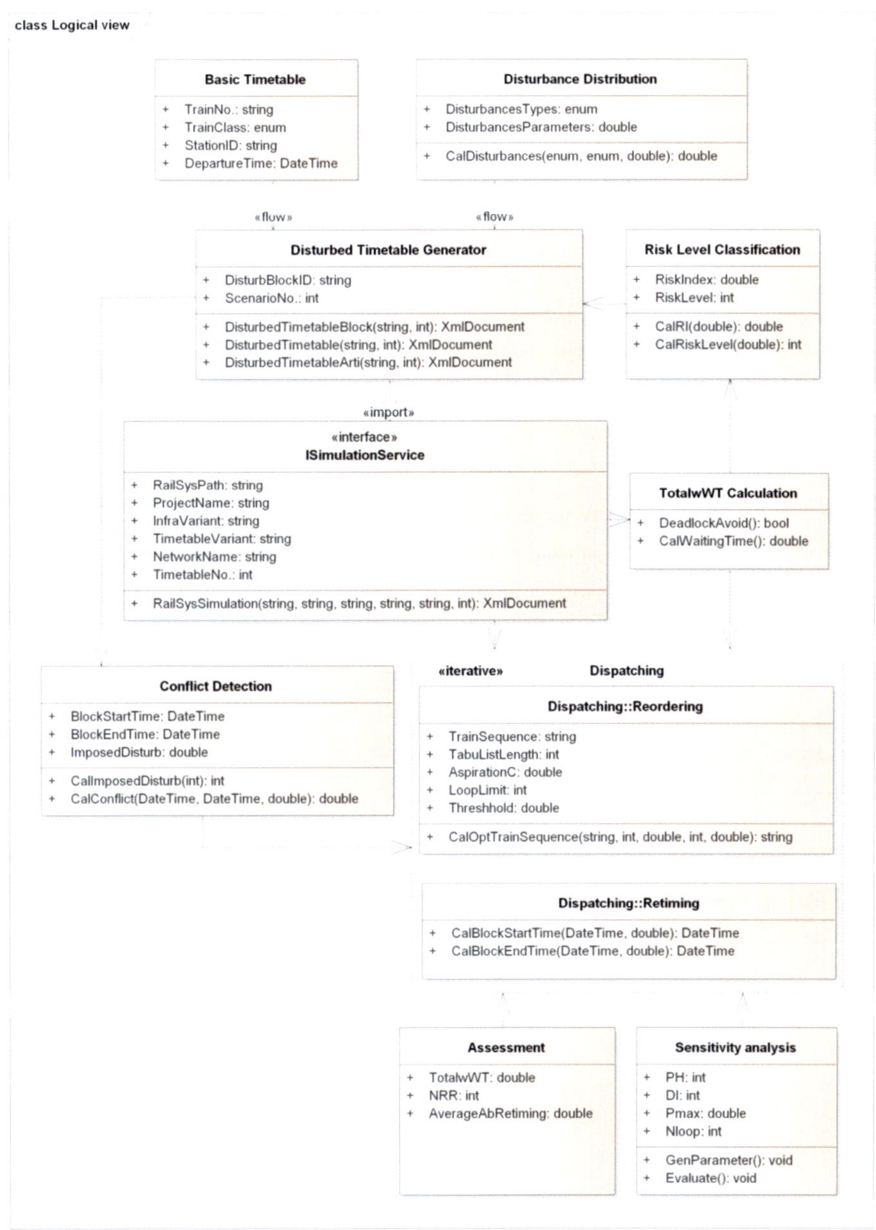

Figure 8-3   Logical view of the dynamic dispatching algorithm

The class "Assessment" evaluates performance of dispatching solutions through 3 perspectives (see Chapter 5), while the class "sensitivity analysis" estimates the effects of the dispatching-related parameters from the generated simulation variants (see Chapter 6). The interface "ISimulationService" standardizes the uniform conditions of different simulation tools. The details of a simulation tool can be isolated from the dispatching system by the data encapsulation in the respective class that implements the interface "ISimulationService".

8.4.2   Process view

The process view describes the behavioral aspects of the dispatching system taking account of the non-functional requirements. In this section, two sequence diagrams are presented to demonstrate the process view of the operational risk classification (see Figure 8-4) and the process view of the proposed dynamic dispatching system (see Figure 8-5). The communication between different classes is specified below.

When the operational risk analysis is carried out, the parameters of the disturbance distribution are first set for generating the disturbance samples. Multiple disturbed timetables are input in the interface "ISimulationService" for calculating operational risk indicators until the mentioned processes implemented iteratively for all the block sections. Finally, the operational risk level obtained based on calculated NRI is returned to the user.

The communication between the interface "ISimulationService" and the client of the interface (Operational Risk Classification) is realized by asynchronous message exchange, since the execution of a simulation would take time. Thus, the operational risk classification system does not have to wait for the simulation results but can process other tasks (generation of disturbance scenarios).

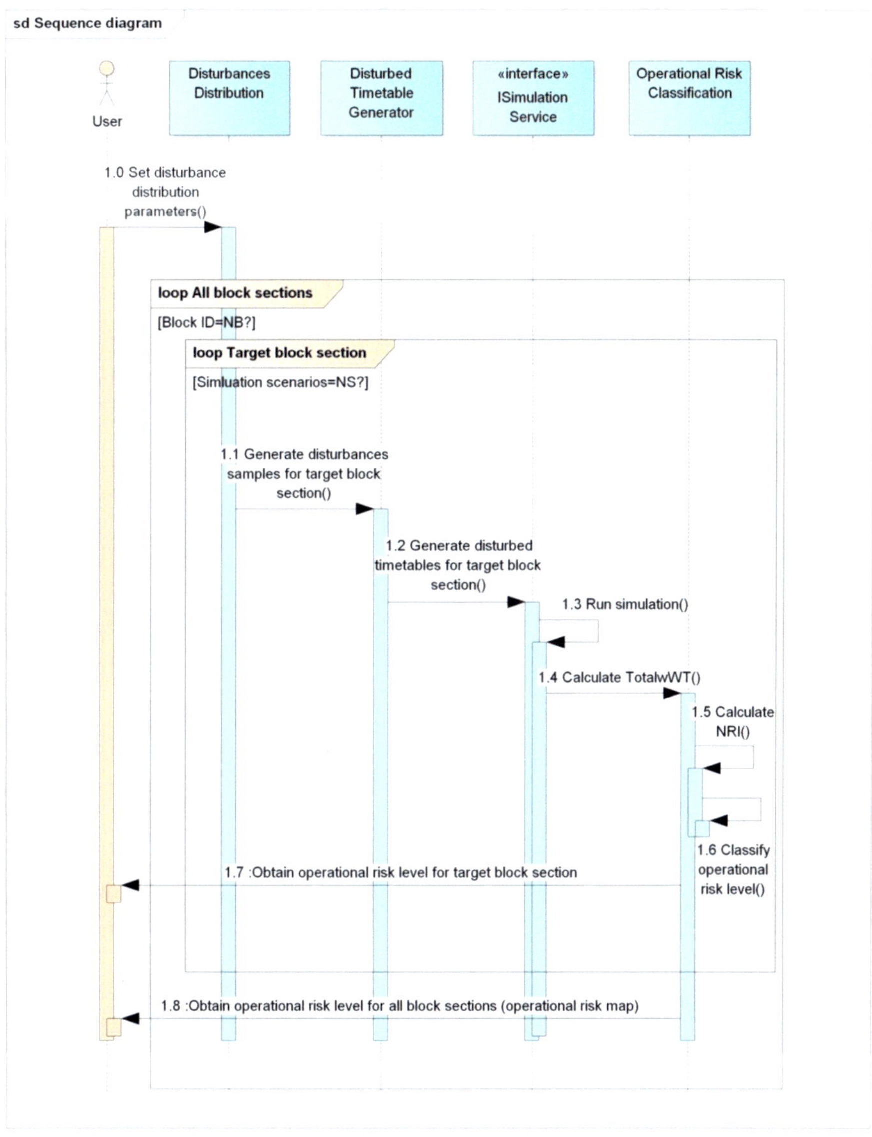

Figure 8-4   Process view of the operational risk classification

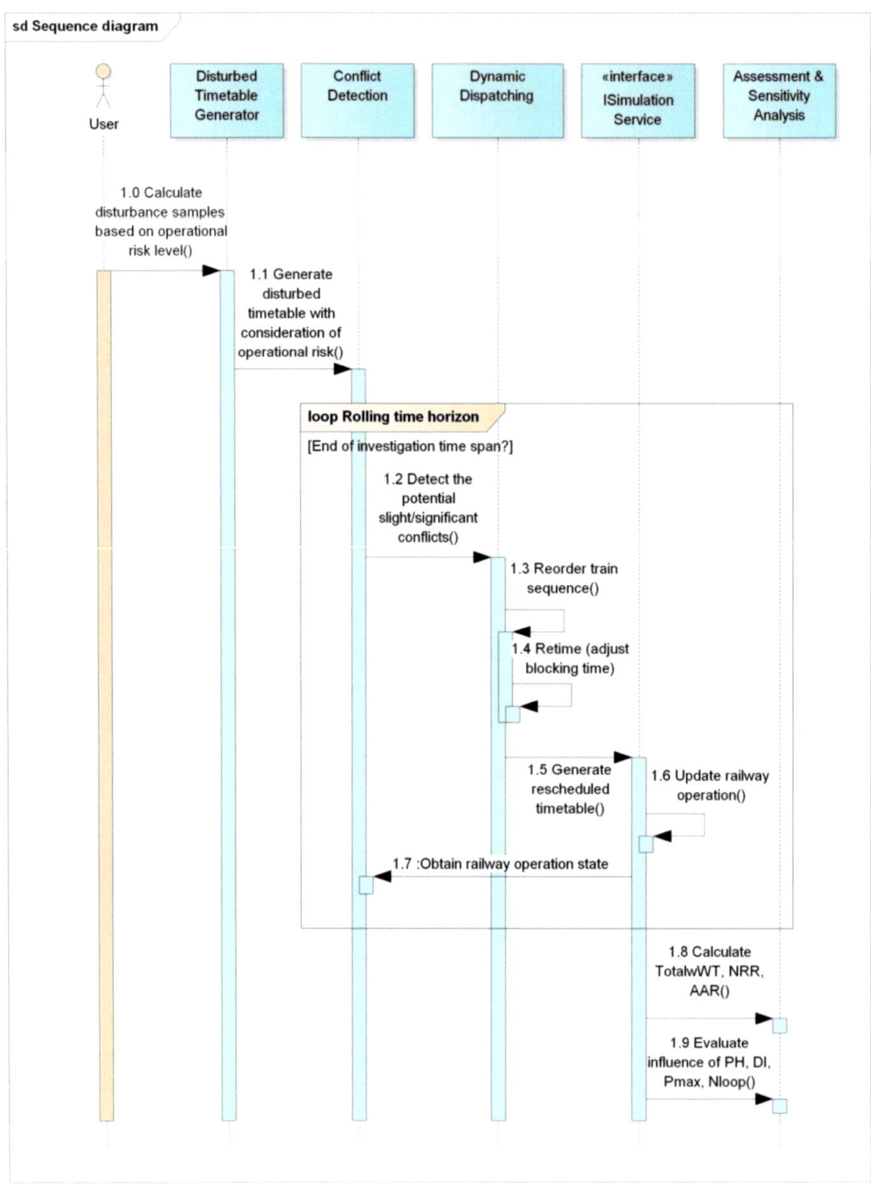

Figure 8-5   Process view of the dynamic dispatching algorithm

Once the operational risk levels of all block sections are confirmed, the class "Disturbed Timetable Generator" is activated. "Conflict Detection" is then conducted based on the disturbed timetable with decisions of consideration amount regarding to potential disturbances. Afterwards, the dynamic dispatching system is implemented by iteratively employing interface "ISimulationService" with rolling time horizon. Assessment as well as sensitivity analysis is fulfilled through calculation of evaluation indicators ($TotalwWT$, NRR, and AAR) and the influence of dispatching-related parameters (PH, DI, P_{max}, and N_{loop}).

Similarly, the communication between the class "Assessment & Sensitivity Analysis" and the interface "ISimulationService" is realized also by asynchronous message exchange. An additional advantage of the asynchronous message exchange is the loose coupling between the dispatching system and the simulation tool. This increases the extensibility and changeability of the implemented dispatching system.

8.4.3 Development view

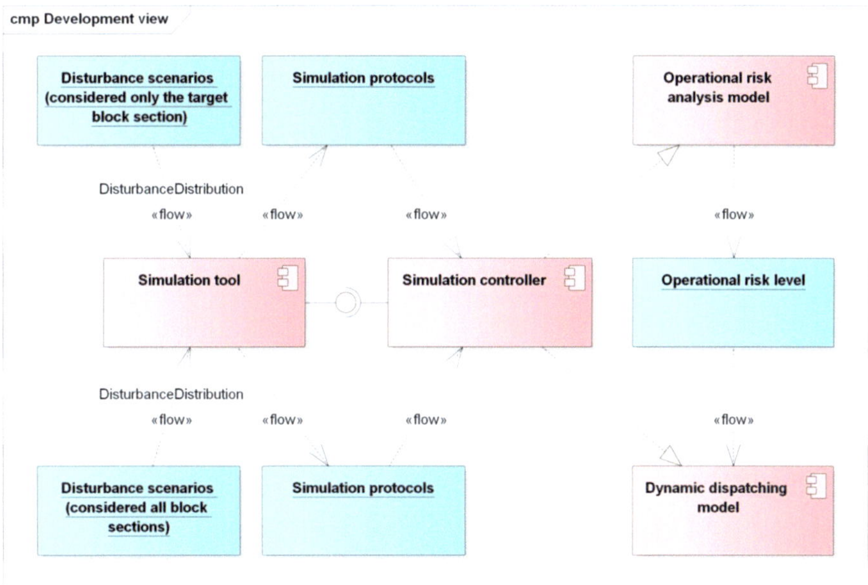

Figure 8-6   Development view of the the dynamic dispatching algorithm

The implementation details of development view are specified through a component diagram in this section (see Figure 8-6).

The dispatching model can be integrated with the various simulation tools via a specified interface through the component "Simulation controller". This makes the internal dispatching system and the external simulation tool independent. Within the scope of this study, a dynamic dispatching system was connected to the simulation tool RailSys®. The simulation controller, which connects to RailSys® via the interface "ISimulationService", controls the communication between the calibration system and RailSys®. If the simulation tool is replaced, the simulation controller can be replaced by another simulation controller. The data flows are represented by dotted lines in the component diagram. The different types of disturbed timetables (see Section 8.4.1) are input in the simulation tool and the specific format simulation protocols are read by the simulation controller. The message exchange is implemented by an internal event observer in the simulation controller component.

9   Conclusions and prospective

The main results of the proposed algorithm are summarized in this chapter. Moreover, the outlooks for potential further research are discussed subsequently.

9.1   Conclusions

The practical significance of this topic stems from the limited expansion and increasing travel demand of railway network. In order to handle the various interferences during the actual railway operation and automatically generate robust dispatching solutions as an assistance for nowadays manuel dispatching, an operational risk-based dynamic dispatching system is developed in this dissertation. Within the framework of this algorithm, the future potential random disturbances are firstly taken into consideration in the dispatching process. The main conclusions are drawn in the following paragraphs.

Aiming at conducting effective dispatching actions under the consideration of future random disturbances, it is necessary to primarily investigate the operational risk of different block sections in the railway network. This study proposes an innovative approach to evaluate the operational risk of block sections based on artificially generated disturbances and railway simulation tools. A case study is conducted on a reference railway network to demonstrate the effectiveness of the proposed approach.

The proposed dispatching algorithm is a hybrid dispatching model which combines both heuristic and simulation approaches. One of the typical heuristic method, Tabu search algorithm, is employed to find a near-optimal dispatching solution in each round of dispatching, and Railsys® will be employed to simulate the railway operation based on the generated dispatching solutions. Different from the conventional dispatching algorithm, a modified blocking time stairway (see Figure 4-2 as an example) is introduced for predicting the potential occupancy conflicts. With continuously updates and adjustments of the railway system, the potential conflicts can be recognized in time and the generated dispatching solutions are able to handle future stochastic disturbances.

The comparison of the results with FCFS principle testify the good performance of the prosed algorithm especially in the circumstance of stochastic disturbances with low computational complexity. In addition, the dispatching-related parameters are also investigated in this dissertation and the most appropriate parameters are scientifically chosen for the reference example network, which enables the sufficient good performance with acceptable computation efforts.

This simulation-based approach is free from the limit of real data, and thus multiple groups of simulation are able to be performed for calculation of the expectation value of the operational risk indicator, for provision the various surrogate of the real-world railway operation, and for implementation of the sensitivity analysis of different related parameters. In such a manner, the statistical reliability of the results can be ensured and thus, the credibility and correctness of the employed simulation tools is significant.

In addition, the proposed algorithm framework is not restricted to the selected probability distribution functions of stochastic disturbances, the specific operating program, the investigated infrastructure network or lines and the considered objective systems of the dispatching optimization. As long as the related data and information is provided, this dynamic dispatching system can be easily employed on any railway network according to different research objective system of specific studies.

Another advantage of this proposed algorithm is that the implementation of this dispatching algorithm liberates dispatchers from burdensome priority calculation tasks. The computation time of this algorithm depends on the scale of the investigation area, the number of disturbances scenarios, the execution time of running a single simulation in the simulation tool, the upper limit of searching loops during TS process as its termination condition, etc.

To conclude, these academic findings are significant for the investigations of railway operation research, for the practical dimensioning of infrastructure territories and for the implementation of dynamic dispatching systems in the form of automation. It can be of interest not only for the scientific community and the railway industry, but even for policy makers and transport authority.

9.2 Prospective

The proposed algorithm can be further investigated or improved regarding to the following aspects, which are also worth being considered from various points of views.

In the module of operational risk analysis, stochastic disturbances according to negative exponential distribution and Erlang distribution are generated in a Monte-Carlo scheme. During the sampling process, only the characteristics of train types are treated differently. The characteristics of the block sections (e.g., Block sections which contain a crossing or a turnout are more prone to come across the electro-mechanic failures compared to other single-line block sections) is neglected due to the limitation of the data collection in this dissertation. Further consideration regarding to infrastructure-related failure would widen the field such as intelligent/smart maintenance of the infrastructure.

Conflicts including occupancy conflicts at track segments, occupancy conflicts at scheduled stops, timetable conflicts, dispatching conflicts are all considered in the module of conflict detection. However, connection conflicts, which arise from the broken coupling between the connecting trains and feeder trains, are not taken into account. It usually requires additional dispatching solutions (e.g., certain rules) to ensure the journey of customers or transport of goods is maintained. For the further research, connection conflicts can also be included in the automatic dispatching system, so as to achieve not only good trains' punctuality but also good passengers' and goods' punctuality.

Similar to the conflict detection procedures, the objectives for guiding the searching solution space and the indicators for evaluating the proposed dispatching system are mainly from the perspectives of railway operating companies and railway infrastructure companies. Nevertheless, it is also meaningful to consider the dispatching solutions from the perspectives of the passengers or customers. Information about the passenger preferences as well as passenger-oriented objective systems can be dug through questionnaires in the future. In such a way of thinking, a better perception of punctuality for passengers can be obtained.

The efficiency of data process is also a major concern of the proposed dispatching algorithm. Except for the parameters stated in Section 9.1, which influence the computation time for the operational risk analysis module and the automatic generation of the dispatching solutions, further studies can be conducted to improve the computation efficiency. For instance, multi-thread of multi-computer can be employed simultaneously for executing the simulation as along as the purchased licenses of the simulation tool are sufficient or the parallel computing technology can be integrated in the framework of Tabu Search so as to accelerate the optimization process.

What's more, the dispatching algorithm employed an external simulation tool for updating the railway operation. Each round of single simulation is launched from an external executable timetable generator, which requires large amount of time for initialization and configuration. In addition, the indicators calculation is achieved through a parser which reads the protocol log-files of each single simulation. The system performance is deteriorated dramatically by a frequent disk *I/O (input/output)* stream. In the future work, the proposed dynamic dispatching can be integrated with the available open source simulation software DoSim or Multi-scaleSim (Liang 2017), so that an improvement of computation time is accomplished.

Appendix

Symbolization of Enterprise Architect (UML)

The UML offers a standard way to visualize, specify and document the system's blueprints. The legend of unified building blocks is given below. Note that the same shape and color stands for one specific type of the building block, which follows the rules of UML models.

A1.   Elements of Use Case

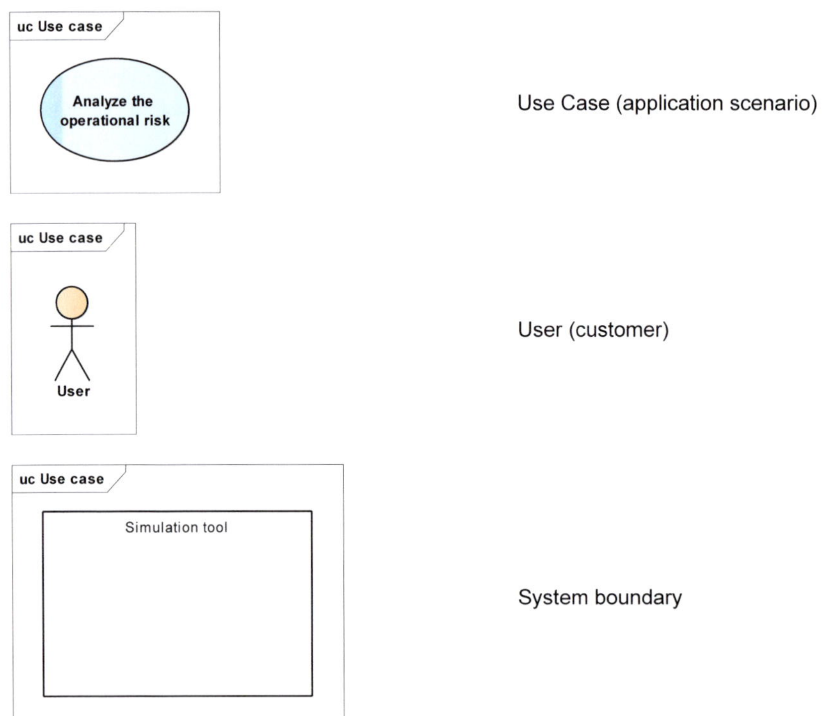

Use Case (application scenario)

User (customer)

System boundary

A2.   Elements of Class Diagram

A set of objects with similar structures and behaviors are summarized as a class. It contains attributes and operations.

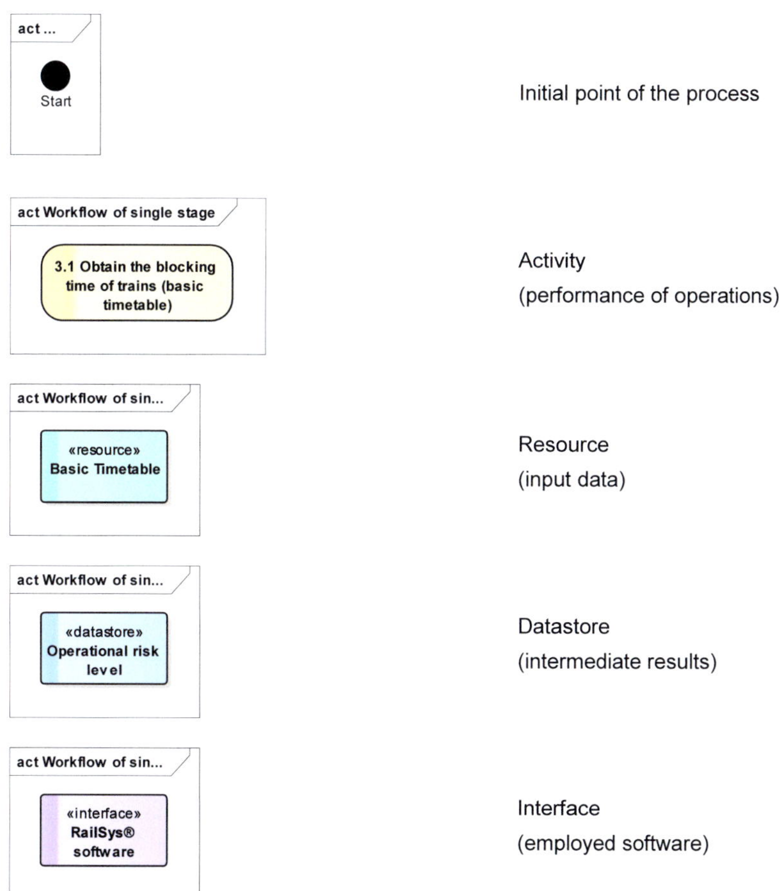

A3. Elements of Activity Diagram

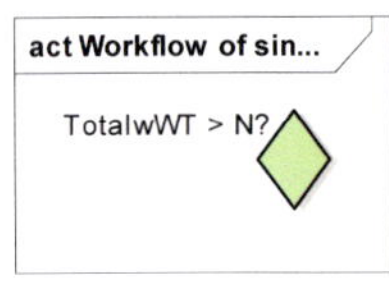

Decision

(different transitions with Boolean conditions)

End point of the process

A4. Elements of Sequence Diagram

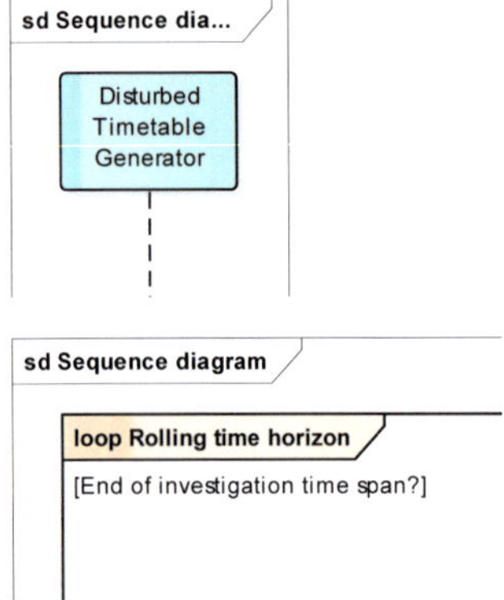

Lifeline of objects

Interaction fragments

(loop, conditional, optional)

A5. Elements of Component Diagram

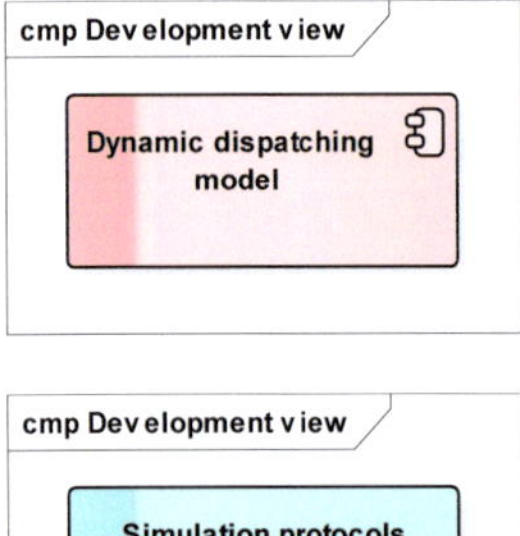

Component

(physical part of the system)

Object

(instance of class including data values)

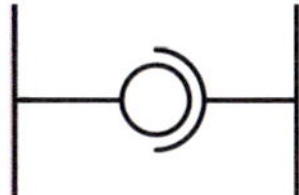

Interface connector

A6.   Elements of Deployment Diagram

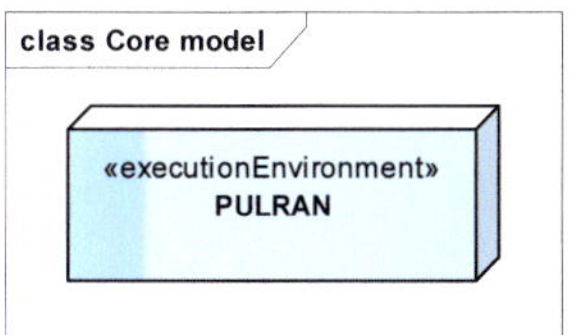

Execution environment

(executable artifacts for specific types of components)

List of Abbreviations

AAR	Average absolute retiming
CDF	Cumulative Distribution Function
DI	Dispatching Interval
DP	Dispatching Point
DSSs	Decision Support Systems
FCFS	First-Come First-Serve
FCLS	First-Come Last-Serve
FRz	Long Distance Passenger Transport
I/O	Input/Output
IEV	Institute of Railway and Transportation Engineering, in German: Institut für Eisenbahn- und Verkehrswesen
NGz	Freight Transport
NRI	Normalized Risk Index
NRR	Number of relative reordering
NRz	Regional Passenger Transport
PDF	Probability Distribution Functions
PH	Prediction Horizon
PULEIV	Program of Capacity Research (investigation of performance), in German: Programm zur Untersuchung des Leistungsverhaltens
PULRAN	Program to Research the Logistic in the Shunting Operation, in German: Programm zur Untersuchung der Logistik im Rangierdienst
RI	Risk Index
ROMA	Railway traffic Optimization by Means of Alternative graphs
SZ	Süddeutsche Zeitung
TotalwWT	Total Weighted Waiting Time
TS	Tabu Search
UIC	International Union of Railway
UML	Unified Modeling Language

List of Symbols

B_h	Block section with high operational risk
B_j	Target block section
$\bar{b}_{t,j}$	Start of blocking time of train t on block section j in the disturbed timetable
$b_{t,j}$	Start of blocking time of train t on block section j in the simulation results
D_{max}	Upper limits of disturbance values which can be generated according to the disturbance distributions
Dep_{basic}	Departure time in basic timetable
Dep_{dis}	Departure time in disturbed timetable
E	Expected loss of the corresponding events
$\bar{e}_{t,j}$	End of blocking time of train t on block section j in the disturbed timetable
$e_{t,j}$	End of blocking time of train t on block section j in the simulation results
$f(S)$	Objective function of Tabu search
$\bar{L}_{cond}$	Mean value of L_i among all condensed basic timetables
L_i	Risk level of a certain block section under the i-th condensed basic timetable
L_{max}	Maximum value of operational risk level
L_{min}	Minimum value of operational risk level
loc	Location reached by train j at time t according to the rescheduled timetable
L_{TL}	Pre-defined length of Tabu list
L_{total}	Total number of operational risk level
M	Total number of trains
N	Total number of block sections on a train path
N_{block}	Number of block sections in the whole investigation area
N_{cond}	Number of condensed basic timetable
N_{loop}	Search loop limit of Tabu search algorithm
NRI_j	Normalized risk index of block section j

NS	Total number of disturbance scenarios
$NStd$	Normalized Std
NT	Threshold of total weighted waiting time to differentiate slight and significant conflicts
P	Probable frequency of uncertain future events
P_{max}	Maximum value of cumulative probability corresponding to the disturbances distribution.
P_{min}	Minimum value of cumulative probability corresponding to the disturbances distribution.
R	Risk
RI_B	Averaged RI of basic timetable
RI_j	Risk index of block section j
RI_{max}	Maximum RI value among all block sections
Run_{basic}	Running time in basic timetable
Run_{dis}	Running time in disturbed timetable
s	Scenario index
S	Potential dispatching solution regarding to Tabu search
S'	Improved dispatching solution regarding to Tabu search
$S*$	Optimal dispatching solution regarding to Tabu search
Std	Standard deviation of operational risk levels among all condensed basic timetable
$Stop_{basic}$	Dwell time in basic timetable
$Stop_{dis}$	Dwell time in disturbed timetable
T_E	Investigation time span
$TotalwWT_s$	Total weighted waiting time of disturbances scenario s
ϑ	Pseudo-random number between 0 and 1
W_e	Proportion parameter which controls the percentage of trains that might be influenced by random disturbances

w_t — Weight of train t for calculating the total weighted waiting time

ΔDep — Produced random samples of departure time extension

ΔDw — Produced random samples of dwell time extension

ΔEn — Produced random samples of entry delay

ΔRun — Produced random samples of running time extension

$\tau_{j,loc}^{TT}(t)$ — Passing time of train j at location *loc* as in the basic timetable: for stations with a stop, this is the time published in the timetable; on block sections, this is the planned time at which the train is expected to pass, following an unhindered path, and considering running time supplements.

Glossary

Some commonly used technical terms are cited from (Hansen, Pachl 2014; Pachl 2002).

Analytical model

A description in terms of mathematical relationships between certain entities those are relevant in a certain problem setting.

Basic structure

The maximum occupation unit allowed to be occupied by only one train simultaneously on a microscopic level without direction information. The boundaries of a basic structure can be the closest signal, signal release point, route release point as well as the borderline of the investigation area.

Basic timetable

A basic timetable includes a set of information with detailed train plans, defining several months in advance the train order and timing at crossings, junctions and platforms.

Block section

A section of track in a fixed block system which a train may enter only when it is not occupied by other vehicles.

Bottleneck relevance

The potential that an infrastructure section will appear as a bottleneck under pre-defined conditions (certain structure of the operating program) in an investigation area.

Bottleneck significance

An indicator for describing whether a bottleneck (depends on the pre-defined limit of operating quality, the structure of a certain operating program and the studied traffic volume) actually has operational influence.

Buffer time

An extra time which is added to the minimum line headway to avoid the transmission of small delays.

Component of train paths

The smallest directional occupancy elements on a train

path which can be divided.

Condensed and disturbed timetable	Timetable generated by imposing disturbances on the condensed basic timetable.
Condensed basic timetable	Timetable generated by compressing the basic timetable to different proportions for compact schedules without changing the mixture of train types.
Consecutive delay	A delay that was transmitted from another train.
Departure time extension	Time extension after the process of passenger boarding and alighting or the freight loading/unloading is finished. It is caused by technical failures of the infrastructure ahead or the trains itself.
Disturbed timetable	Timetable generated by imposing disturbances on the basic timetable.
Dwell time extension	Unscheduled time extension between the scheduled dwell time and the actual dwell time at each scheduled stop. For passenger train types, it is mainly caused by the time extension of passenger boarding and alighting. For freight train types, it is related to the unloading or loading time.
Entry delay	Difference between the scheduled arrival time and actual arrival time of involved trains at the boundary of the investigated area.
Objective function	A mathematical representation of the objective that is aimed at in terms of the decision variables.
Online dispatching	Adjustment of train schedules to the real time ever-changing conditions under the circumstances of delays.
Operating program	A comprehensive data-related description of operations and the traffic units involved in these operations.

	Important components can be: amount, structure, sequence, characteristics of trains, and the temporal allocation of train runs.
Operational disturbances	Collective name includes entry delays, dwell time extensions, running time extensions and departure time extensions.
Optimal solution	A feasible solution to an optimization problem is optimal if another feasible solution with a better objective value does not exist.
Primary delay	A delay that was not transmitted from another train.
Recovery time	A time supplement that is added to the pure running time to enable a train to make up small delays.
Risk index	Indicator for evaluation the operational risk analysis. It is calculated as the average value of the total weighted waiting time among all disturbances scenarios.
Robustness of dispatching	A dispatching solution efficient for a larger immediate time span without incurring repetitive dispatching action.
Robustness of timetable	A timetable can handle small disturbances which occur in real operation, also known as low sensitivity to disturbances with random variables.
Rolling time horizon	An optimization framework which solves dynamic problems into sub-static-problems iteratively. It is subject to handle uncertainties in the system by conducting actions per control horizon. In order to make appropriate decisions, consecutive prediction horizons within the framework are overlapped.

Running time extension   Unscheduled extension of running time caused by the disturbances during train-running or the stochastic behavior of the train driver.

Simulation model   A mathematical representation of a system that can be used to mimic the processes of the system under varying circumstances. Usually, a simulation model is operated subject to stochastic disturbances.

Tabu Search   A metaheuristic search model which employs neighborhood search with memory structures applied for optimization problems.

Total weighted waiting time   Weighted sum of each train's unscheduled waiting time.

Turnout   An assembly of rails, movable points, and a frog, which effects the tangential branching of tracks and allows trains or vehicles to run over one track or another.

Unscheduled waiting time   Waiting time caused by hindrance during operational process, which is not included in the timetable.

Publication bibliography

Aboudolas, K.; Papageorgiou, M.; Kouvelas, A.; Kosmatopoulos, E. (2010): A rolling-horizon quadratic-programming approach to the signal control problem in large-scale congested urban road networks. In *Transportation Research Part C: Emerging Technologies* 18 (5), pp. 680–694.

Albrecht, A.; Panton, D.; Lee, D. (2013): Rescheduling rail networks with maintenance disruptions using Problem Space Search. In *Computers & Operations Research* 40 (3), pp. 703–712.

Allan, J.; Brebbia, C.; Hill, R.; Sciutto, G.; Sone, S. (Eds.) (2004): Computers in Railways IX. Southampton: WIT Press.

Andersson, E.; Peterson, A.; Törnquist Krasemann, J. (2013): Quantifying railway timetable robustness in critical points. In *Journal of Rail Transport Planning & Management* 3 (3), pp. 95–110.

Bidot, J. (2003): Using Constraint Programming and Simulation for Execution Monitoring and On-Line Rescheduling with Uncertainty. In Gerhard Goos, Juris Hartmanis, Jan van Leeuwen, Francesca Rossi (Eds.): Principles and Practice of Constraint Programming – CP 2003, vol. 2833. Berlin, Heidelberg: Springer Berlin Heidelberg (Lecture Notes in Computer Science), p. 966.

Boccia, M.; Mannino, C.; Vasilyev, I. (2013): The dispatching problem on multitrack territories. Heuristic approaches based on mixed integer linear programming. In *NETWORKS* 62 (4), pp. 315–326.

Brännlund, U.; Lindberg, P.; Nõu, A.; Nilsson, J.-E. (1998): Railway Timetabling Using Lagrangian Relaxation. In *Transportation Science* 32 (4), pp. 358–369.

Briggs, K.; Beck, C. (2007): Modelling train delays with q-exponential functions. In *Physica A: Statistical Mechanics and its Applications* 378 (2), pp. 498–504.

Bruinsma, F.; Rietveld, P.; van Vuuren, D. (Eds.) (1999): Unreliability in public transport Chains. Selected Proceedings of the 8th World Conference on Transport Research. Rotterdam: Tinbergen Inst (Discussion paper / Tinbergen Institute 3).

Bussieck, M.; Kreuzer, P.; Zimmermann, U. (1997): Optimal lines for railway systems. In *European Journal of Operational Research* 96 (1), pp. 54–63.

Caimi, G.; Fuchsberger, M.; Laumanns, M.; Lüthi, M. (2012): A model predictive control approach for discrete-time rescheduling in complex central railway station areas. In *Computers & Operations Research* 39 (11), pp. 2578–2593.

Carey, M.; Carville, S. (2000): Testing schedule performance and reliability for train stations. In *The Journal of the Operational Research Society* 51 (6), p. 666.

Carey, M.; Kwieciński, A. (1994): Stochastic approximation to the effects of headways on knock-on delays of trains. In *Transportation Research Part B* 28 (4), pp. 251–267.

Chen, M.; Niu, H. (2013): Optimizing Schedules of Rail Train Circulations by Tabu Search Algorithm. In *Mathematical Problems in Engineering* 2013 (12), pp. 1–7.

Cheng, Y. (1998): Hybrid simulation for resolving resource conflicts in train traffic rescheduling. In *Computers in Industry* 35 (3), pp. 233–246.

Chiu, C.; Chou, C.; Lee, J.; Leung, H.; Leung, Y. (2002): A Constraint-Based interactive train rescheduling tool. In *Constraints* 7 (2), pp. 167–198.

Corman, F.; D'Ariano, A.; Pacciarelli, D.; Pranzo, M. (2010a): A tabu search algorithm for rerouting trains during rail operations. In *Transportation Research Part B: Methodological* 44 (1), pp. 175–192.

Corman, F.; D'Ariano, A.; Pacciarelli, D.; Pranzo, M. (2010b): Centralized versus distributed systems to reschedule trains in two dispatching areas. In *Public Transport* 2 (3), pp. 219–247.

Corman, F.; D'Ariano, A.; Pranzo, M.; Hansen, I. (2011): Effectiveness of dynamic reordering and rerouting of trains in a complicated and densely occupied station area. In *Transportation Planning and Technology* 34 (4), pp. 341–362.

Cui, Y.; Martin, U. (2011): Multi-scale simulation in railway planning and operation. In *Traffic and Transportation* 23 (06), pp. 511–517.

Cui, Y.; Martin, U.; Zhao, W. (2016): Calibration of disturbance parameters in railway operational simulation based on reinforcement learning. In *Journal of Rail Transport Planning & Management* 6 (1), pp. 1–12.

D'Ariano, A.; Pacciarelli, D.; Pranzo, M. (2007): A branch and bound algorithm for scheduling trains in a railway network. In *European Journal of Operational Research* 183 (2), pp. 643–657.

D'Ariano, A. (2008): Improving real-time train dispatching: models, algorithms and applications. Dissertation. TU Delft, Delft.

D'Ariano, A.; Corman, F.; Hansen, I. (2008): Evaluating the performance of an advanced train dispatching system. In : 2008 First International Conference on Infrastructure Systems and Services: Building Networks for a Brighter Future (INFRA). 2008 First International Conference on Infrastructure Systems and Services: Building Networks for a Brighter Future (INFRA). Rotterdam, Netherlands, 10.11.2008 - 12.11.2008: IEEE, pp. 1–6.

DB NETZ AG (2008): DB Richtlinie 405 - Fahrwegkapazität.

DB NETZ AG (2010): DB Richtlinie 420.02.

DB NETZ AG (2015): The train path pricing system of DB Netz AG.

Dewilde, T.; Sels, P.; Cattrysse, D.; Vansteenwegen, P. (Eds.) (2011): Defining robustness of a railway timetable. 4th International Seminar on Railway Operations Modelling and Analysis (IAROR) : RailRome 2011, Proceedings. Rome, Italy, 2011-02-16. Ghent, Belgium: Ghent University, Department of Industrial Management.

Dorfman, M.; Medanic, J. (2004): Scheduling trains on a railway network using a discrete event model of railway traffic. In *Transportation Research Part B: Methodological* 38 (1), pp. 81–98.

Drewello, H.; Günther, F. (Eds.) (2012): Bottleneck in railway infrastructure - do they really exist? The corridor Rotterdam - Genoa. European transport conference. Glasgow.

Dündar, S.; Şahin, İ. (2013): Train re-scheduling with genetic algorithms and artificial neural networks for single-track railways. In *Transportation Research Part C: Emerging Technologies* 27, pp. 1–15.

Espinosa-Aranda, J.; García-Ródenas, R. (2012): A Discrete Event-Based Simulation Model for Real-Time Traffic Management in Railways. In *Journal of Intelligent Transportation Systems* 16 (2), pp. 94–107.

Freund, J.; Jones, J. (2014): Measuring and Managing Information Risk. A FAIR Approach. Burlington: Elsevier Science.

Geiger, W. (1994): Qualitäts Lehre: Einführung · Systematik · Terminologie. 2. Auflage. Braunschweig/Wiesbaden: Vieweg Verlag.

Girish, B. (2016): An efficient hybrid particle swarm optimization algorithm in a rolling horizon framework for the aircraft landing problem. In *Applied Soft Computing* 44, pp. 200–221.

Glover, F. (1986): Future paths for integer programming and links to artificial intelligence. In *Computers & Operations Research* 13 (5), pp. 533–549.

Glover, F. (1989): Tabu Search—Part I. In *ORSA Journal on Computing* 1 (3), pp. 190–206.

Glover, F. (1990): Tabu Search—Part II. In *ORSA Journal on Computing* 2 (1), pp. 4–32.

Gorman, M. (1998): An application of genetic and tabu searches to the freight railroad operating plan problem. In *Annals of Operations Research* 78, pp. 51–69.

Goverde, R. (2005): Punctuality of railway operations and timetable stability analysis. Delft: TRAIL (TRAIL thesis series, T2005/10).

Goverde, R. (2007): Railway timetable stability analysis using max-plus system theory. In *Transportation Research Part B: Methodological* 41 (2), pp. 179–201.

Goverde, R. (2010): A delay propagation algorithm for large-scale railway traffic networks. In *Transportation Research Part C: Emerging Technologies* 18 (3), pp. 269–287.

Goverde, R.; Hooghiemstra, G.; Lopuhaä, H. (2001): Statistical analysis of train traffic. The Eindhoven case. Delft, Netherlands: DUP Science (TRAIL studies in transportation science, no. S2001/1).

Grube, P.; Núñez, F.; Cipriano, A. (2011): An event-driven simulator for multi-line metro systems and its application to Santiago de Chile metropolitan rail network. In *Simulation Modelling Practice and Theory* 19 (1), pp. 393–405.

Hansen, I.; Pachl, J. (2014): Railway timetabling and operations. Analysis, modelling and simulation. 2nd ed. Hamburg: Eurailpress.

Hantsch, F.; Li, X.; Martin, U. (2013): Methoden zur Engpassanalyse bei der Infrastrukturbemessung im Schienenverkehr. In *ETR* 62 (3), pp. 30–33.

Hartwig, R. (2013): Bottleneck analysis. South East Transport Axis.

Hellström, P. (1998): Analysis and evaluation of systems and algorithms for computer-aided train dispatching. Uppsala (UPTEC, 98 008R).

Higgins, A.; Kozan, E. (1998): Modeling Train Delays in Urban Networks. In *Transportation Science* 32 (4), pp. 346–357.

Higgins, A.; Kozan, E.; Ferreira, L. (1997): Heuristic Techniques for Single Line Train Scheduling. In *Journal of Heuristics* 3 (1), pp. 43–62.

Huisman, D.; Kroon, L.; Maróti, G. (2007): Railway timetabling from an operations research perspective. Rotterdam: Econometric Institute (Econometric Institute report EI, 22).

Hutchison, D.; Kanade, T.; Kittler, J.; Kleinberg, J.; Mattern, F.; Mitchell, J. et al. (Eds.) (2009): Robust and Online Large-Scale Optimization. Berlin, Heidelberg: Springer Berlin Heidelberg (Lecture Notes in Computer Science).

IEV (2005): Spezifikation Zuglaufregelung (not published). Institut für Eisenbahn- und Verkehrswesen. Universität Stuttgart.

Iyer, R.; Ghosh, S. (1995): DARYN-a distributed decision-making algorithm for railway networks. Modeling and simulation. In *IEEE Trans. Veh. Technol.* 44 (1), pp. 180–191.

Kacprzyk, J.; Xhafa, F.; Abraham, A. (Eds.) (2008): Metaheuristics for Scheduling in Industrial and Manufacturing Applications. Berlin, Heidelberg: Springer Berlin Heidelberg (Studies in Computational Intelligence).

Keaton, M. (1989): Designing optimal railroad operating plans. Lagrangian relaxation and heuristic approaches. In *Transportation Research Part B: Methodological* 23 (6), pp. 415–431.

Koch, W. (Ed.) (2000): A low cost tool for rescheduling in regional public transport systems. Proceedings of the 9th IFAC Symposium on Contol in Transport Systems. Braunschweig.

Kroon, L.; Maróti, G.; Helmrich, M.; Vromans, M.; Dekker, R. (2008): Stochastic improvement of cyclic railway timetables. In *Transportation Research Part B: Methodological* 42 (6), pp. 553–570.

Kruchten, P. (1995): Architectural blueprints - the "4+1" view model of software architecture. In *IEEE Software* 12 (6), pp. 42–50.

Lamma, E.; Mello, P.; Milano, M. (1997): A distributed constraint-based scheduler. In *Artificial Intelligence in Engineering* 11 (2), pp. 91–105.

Larroche, Y.; Moulin, R.; Gauyacq, D. (1996): SEPIA. A real-time expert system that automates train route management. In *Control Engineering Practice* 4 (1), pp. 27–34.

Lee, T.; Ghosh, S. (2001): Stability of RYNSORD-a decentralized algorithm for railway networks under perturbations. In *IEEE Trans. Veh. Technol.* 50 (1), pp. 287–301.

Li, X.; Martin, U. (2015): Ursachenbezogene Engpassbewertung in der Eisenbahnbetriebssimulation. In *ETR* 64 (1+2), pp. 40–45.

Li Ping; Nie Axin; Jia Limin; Wang Fuzhang (Eds.) (2001): Study on intelligent train dispatching. Proceedings of IEEE Intelligent Transportation Systems Conference. Oakland, CA, USA, 25-29 Aug.

Liang, J. (2017): Metaheuristic-based dispatching optimization integrated in multi-scale simulation model of railway operation. Dissertation. University of Stuttgart, Stuttgart.

Lindfeldt, A. (2015): Railway capacity analysis. Methods for simulation and evaluation of timetables, delays and infrastructure. Stockholm, [Sweden]: KTH Royal Institute of Technology, School of Architecture and the Built Environment, Department of Transport Science (TRITA-TSC-PHD, 15-002).

Lüthi, M. (2009): Improving the efficiency of heavil used railway networks through integrated real-time rescheduling. Ph.D Thesis. ETH Zurich.

Lüthi, M.; Nash, A.; Weidmann, U.; Laube, F.; Wuest, R. (Eds.) (2007): Increasing railway capacity and reliability through integrated real-time rescheduling. Proceedings of the 11th world conference on transport research. Berkeley.

Martin, U. (1995): Verfahren zur Bewertung von Zug- und Rangierfahrten bei der Disposition. TU Braunschweig. Braunschweig: Inst. für Eisenbahnwesen und Verkehrssicherung (Schriftenreihe / Institut für Verkehr, Eisenbahnwesen und Verkehrssicherung, Technische Universität Braunschweig, 52).

Martin, U. (2014a): Anforderungsgerechte Systematik zur Gestaltung von Zeitzuschlägen im Bahnbetrieb (ATRANS 1). DFG-Projekt mit Nummer "MA 2326/14-1". Institut für Eisenbahn- und Verkehrswesen. Stuttgart.

Martin, U. (2014b): The influence of dispatching on the relationship between capacity and operation quality of railway systems. DFG-Project with project number "MA 2326/15-1". Institut für Eisenbahn- und Verkehrswesen. Stuttgart.

Martin, U.; Chu, Z. (2014): Direkte experimentelle Bestimmung der maximalen Leistungsfähigkeit bei Leistungsuntersuchungen im spurgeführten Verkehr. DFG-Forschungsprojekt (MA 2326/6-1). 1. Aufl. Norderstedt: Books on Demand (Neues verkehrswissenschaftliches Journal, 7).

Martin, U.; Li, X. (2015): Entwicklung einer simulationsbasierten Methodik zur ursachenbezogenen Engpassbewertung komplexer Gleisstrukturen in spurgeführten Verkehrssystemen unter Berücksichtigung stochastischer Bedingungen. DFG-Forschungsprojekt (MA 2326/10-1). Norderstedt: Books on Demand (Neues verkehrswissenschaftliches Journal, 11).

Martin, U.; Li, X.; Schmidt, C. (2008): PULEIV Projektbericht: Allgemeingültiges Verfahren zur praxisorientierten Bestimmung des Leistungsverhaltens von Eisenbahninfrastrukturen. Im Auftrag der DB Netz AG.

Martin, U.; Li, X.; Warninghoff, C.-R. (2012): Bewertungsverfahren für Knotenelemente bei der Infrastrukturbemessung - RePlan. In *ETR* 11, pp. 38–43.

Martin, U.; Liang, J.; Cui, Y. (2015): Einfluss von ausgewählten Dispositionsparametern auf das Ergebnis von Leistungsuntersuchungen. In *ETR* 64 (7+8), pp. 20–23.

Meng, L.; Zhou, X. (2011): Robust single-track train dispatching model under a dynamic and stochastic environment. A scenario-based rolling horizon solution approach. In *Transportation Research Part B: Methodological* 45 (7), pp. 1080–1102.

Nielsen, L.; Kroon, L.; Maróti, G. (2012): A rolling horizon approach for disruption management of railway rolling stock. In *European Journal of Operational Research* 220 (2), pp. 496–509.

Nießen, N. (2008): Leistungskenngrößen für Gesamtfahrstraßenknoten. Ph.D. Thesis, Aachen. RWTH Aachen University.

Oetting, A. (2013): Experiences of optimising heavily used mixed-traffic lines. IT13 Rail. Zürich, 2013.

OpenTrack. Railway Simulation: OpenTrack Railway Technology Ltd. Available online at http://www.opentrack.ch/opentrack/opentrack_d/opentrack_d.html.

Pachl, J. (2002): Railway Operation and Control: VTD Rail Pub.

Pänke, I.; Klimmt, A. (2012): Vernetzung von Technologie und Anwenderwissen. In *Deine Bahn* 3, pp. 44–47.

Peeta, S.; Mahmassani, H. (1995): Multiple user classes real-time traffic assignment for online operations: A rolling horizon solution framework. In *Transportation Research Part C: Emerging Technologies* 3 (2), pp. 83–98.

Plöchinger, S.; Jaschensky, W. (2013): So verspätet ist die Bahn in Ihrer Stadt, auf Ihrer Strecke. In *Süddeutsche Zeitung*, 4/8/2013.

Pöhle, D.; Feil, M. (2016): What Are New Insights Using Optimized Train Path Assignment for the Development of Railway Infrastructure? In Marco Lübbecke, Arie Koster, Peter Letmathe, Reinhard Madlener, Britta Peis, Grit Walther (Eds.): Operations Research Proceedings 2014. Cham: Springer International Publishing (Operations Research Proceedings), pp. 457–463.

Pudney, P.; Wardrop, A. (2008): Generating Train Plans with Problem Space Search. In Mark Hickman, Pitu Mirchandani, Stefan Voß (Eds.): Computer-aided Systems in Public

Transport, vol. 600. Berlin, Heidelberg: Springer Berlin Heidelberg (Lecture Notes in Economics and Mathematical Systems), pp. 195–207.

Quaglietta, E.; Corman, F.; Goverde, R. (2013): Stability of railway dispatching solutions under a stochastic and dynamic environment. In *Journal of Rail Transport Planning & Management* 3, pp. 137–149.

RMCON Rail Management Consultants (2010): RailSys. Hannover: RMCON. Available online at http://www.rmcon.de/.

Salido, M.; Barber, F.; Ingolotti, L. (Eds.) (2008): Robustness in railway transportation scheduling. 7th World Congress on Intelligent Control and Automation. Chongqing, China, 25-27 June 2008.

Salim, V.; Cai, X. (1997): A genetic algorithm for railway scheduling with environmental considerations. In *Environmental Modelling & Software* 12 (4), pp. 301–309.

Sayarshad, H.; Ghoseiri, K. (2009): A simulated annealing approach for the multi-periodic rail-car fleet sizing problem. In *Computers & Operations Research* 36 (6), pp. 1789–1799.

Schaer, T.; Jacobs, J.; Scholl, S. und Kurby, S. (2005): DisKon-Laborversion eines flexiblen, modularen und automatischen Dispositionsassistenzsystems. In *ETR* 12, pp. 809–821.

Schlaich, J. (2002): Beitrag zur Entwicklung eines Dispositionswerkzeuges zur Optimierung betriebsbedingter Wartezeiten im schienengebundenen Verkehr. Studienarbeit. Universität Stuttgart, Stuttgart.

Schöbel, A.; Kratz, A. (2009): A Bicriteria Approach for Robust Timetabling. In David Hutchison, Takeo Kanade, Josef Kittler, Jon M. Kleinberg, Friedemann Mattern, John C. Mitchell et al. (Eds.): Robust and Online Large-Scale Optimization, vol. 5868. Berlin, Heidelberg: Springer Berlin Heidelberg (Lecture Notes in Computer Science), pp. 119–144.

Schwanhäusser, W. (1974): Die Bemessung der Pufferzeiten im Fahrplangefüge der Eisenbahn. Ph.D Thesis, Aachen. Veröffentlichungen des Verkehrswissenschaftlichen Institutes der Rheinischen-Westfälischen Technischen Hochschule Aachen.

Shafia, M.; Pourseyed Aghaee, M.; Sadjadi, S.; Jamili, A. (2012): Robust Train Timetabling Problem. Mathematical Model and Branch and Bound Algorithm. In *IEEE Trans. Intell. Transport. Syst.* 13 (1), pp. 307–317.

Sterzik, S.; Kopfer, H. (2013): A Tabu Search Heuristic for the Inland Container Transportation Problem. In *Computers & Operations Research* 40 (4), pp. 953–962.

Takuchi, Y.; TOMII, N.; HIRAI, C. (2007): Evaluation Method of Robustness for Train Schedules. In *QR of RTRI* 48 (4), pp. 197–201.

Törnquist, J. (2012): Design of an effective algorithm for fast response to the re-scheduling of railway traffic during disturbances. In *Transportation Research Part C: Emerging Technologies* 20 (1), pp. 62–78.

Törnquist, J.; Persson, J. (2005): Train Traffic Deviation Handling Using Tabu Search and Simulated Annealing. Los Alamitos, USA (Proceedings of the 38th Annual Hawaii International Conference on System Sciences).

Törnquist, J.; Persson, J. (2007): N-tracked railway traffic re-scheduling during disturbances. In *Transportation Research Part B: Methodological* 41 (3), pp. 342–362.

UIC: UIC Leaflet 406.

Wegele, S.; Schneider, E. (Eds.) (2004): Dispatching of train operations using genetic algorithms. Proceedings of the 9th International Conference on Computer-Aided Scheduling of Public Transport (CASPT). San Diego, USA.

Wegele, S.; Slovak, R.; Schnieder, E. (Eds.) (2004): Automatic dispatching of train operations using a hybrid optimisation method. Proceedings of the 9th International Conference on Computer-Aided Scheduling of Public Transport. San Diego, USA.

Yuan, J. (2006): Stochastic modelling of train delays and delay propagation in stations. Delft: Eburon Academic Publishers; The Netherlands TRAIL Research School] (TRAIL thesis series, T2006/6).

Yuan, J.; Goverde, R.; Hansen, I. (Eds.) (2002): Propagation Of Train Delays In Stations. Southampton: WIT Press (Computers in Railways VIII).

Zhang, Q.; Manier, H.; Manier, M.-A. (2012): A genetic algorithm with tabu search procedure for flexible job shop scheduling with transportation constraints and bounded processing times. In *Computers & Operations Research* 39 (7), pp. 1713–1723.

Zufferey, N.; Verma, M. (2011): Tabu search for shipping dangerous goods in a rail-truck network. In *Research in Logistics and Production* 1 (3), pp. 127–137.